NOISE

CONTROL

SOLUTIONS

FOR THE

FOOD PRODUCTS
INDUSTRY

By Richard K. Miller, Wayne V. Montone,
and Mark D. Oviatt

NOISE CONTROL SOLUTIONS FOR THE
FOOD PRODUCTS INDUSTRY

ISBN: 0-89671-003-3

Published by

The Fairmont Press, Inc.
P.O. Box 14227
Atlanta, Georgia 30324

TABLE OF CONTENTS

TABLE OF CONTENTS (Continued)   *Page*

ABOUT THE AUTHORS

This report was developed by Richard K. Miller & Associates, Inc.,
Consultants in Acoustics, 464 Armour Circle, N. E., Atlanta, Ga.,
30324. Established in 1972, the firm is the oldest and largest in
the Southeast providing consulting services exclusively in the areas
of industrial noise control, architectural acoustics, and environmental noise.

The firm has extensive experience in administering noise control
programs in the food products industry for clients such as Coca-
Cola, Gold Kist, Shasta, Pillsbury, Southern States, Wilson Foods,
Sanitary Dairy, E. Kahn's Son, Derst Baking, American Homes Foods,
Jewel Food, C-B Foods, Pepsi-Cola, and Quaker Oats.

INTRODUCTION

The approaches to noise control which are presented in this report are designed to address noise problems in the food products industry in a specific
manner. It should be recognized, however, that machinery usage will vary
from plant to plant. While the general approaches to noise reduction presented in this report should be applicable to a wide variety of plants,
careful engineering judgement should be made for each potential application
to insure acoustical, production, and safety constraints are considered and
dealt with.

1 - GENERAL APPROACHES TO NOISE CONTROL

Three approaches to noise control should be considered for any noise problem:

1. The noise *source* may be modified.

2. Noise may be blocked or reduced along the *path* from the source to the receiver.

3. Sound may be isolated from the *receiver* by means of barriers, operator location, or hearing protection.

The optimum approach for any operation must be determined based on acoustical effectiveness, production compatibility and economics. It should be pointed out that OSHA recognizes hearing protective devices as only a temporary solution to noise exposure, and stipulates that other engineering methods must be employed as permanent compliance measures.

The first step in reducing noise is to define specifically how the acoustic energy is being generated. All noise sources generate sound by one of the following two mechanisms:

1. Acoustical radiation from a vibrating surface.

2. Aerodynamic turbulence.

Six types of noise control systems may be considered to solve any noise problem:

1. Sound barriers.

2. Sound absorbers.

3. Vibration damping.

4. Vibration isolation.

5. Mufflers.

6. Machine redesign, process modification, or noise source elimination.

Each of these six conceptual approaches is considered in the noise control solutions for specific items of machinery discussed in this report.

The Williams-Steiger Occupational Safety and Health Act of 1970 (Public Law 91-596) was established "to assure safe and healthful working conditions for working men and women...." The Occupational Safety and Health Administration (OSHA) of the U.S. Department of Labor is delegated the responsibility of implementing and enforcing the law.

Title 29 CFR, Section 1910.95 promulgates regulations for the protection of employees from potentially dangerous noise exposure. A copy of the section is presented in Figure 2.1. Proposed revisions to this regulation were published in the Federal Register of October 24, 1974. This revision is still under consideration.

While the OSHA regulations establish a maximum noise level of 90 dBA for a continuous 8 hour exposure during a working day, higher sound levels are allowed for shorter exposure times. Thus, for cyclic operations, it is necessary to compute the employee's noise dose, or percent allowable exposure for actual operation.

Example:

A machine generates sound levels of 95 dBA for 1 minute during each cycle, 200 times per day. From Figure 2.1, the operator's daily noise dose is:

$$D = \frac{C}{T} = \frac{200 \text{ minutes}}{4 \text{ hours}} = \frac{3.33}{4} = 83\%$$

This dosage is within the OSHA limit of 100%.

§ 1910.95 Occupational noise exposure.

(a) Protection against the effects of noise exposure shall be provided when the sound levels exceed those shown in Table G-16 when measured on the A scale of a standard sound level meter at slow response. When noise levels are determined by octave band analysis, the equivalent A-weighted sound level may be determined as follows:

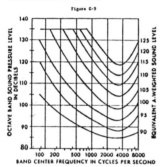

Figure G-9

Equivalent sound level contours. Octave band sound pressure levels may be converted to the equivalent A-weighted sound level by plotting them on this graph and noting the A-weighted sound level corresponding to the point of highest penetration into the sound level contours. This equivalent A-weighted sound level, which may differ from the actual A-weighted sound level of the noise, is used to determine exposure limits from Table G-16.

[1910.95 amended at 39 FR 19468, June 3, 1974]

(b) (1) When employees are subjected to sound exceeding those listed in Table G-16, feasible administrative or engineering controls shall be utilized. If such controls fail to reduce sound levels within the levels of Table G-16, personal protective equipment shall be provided and used to reduce sound levels within the levels of the table.

(2) If the variations in noise level involve maxima at intervals of 1 second or less, it is to be considered continuous.

(3) In all cases where the sound levels exceed the values shown herein, a continuing, effective hearing conservation program shall be administered.

TABLE G-16—PERMISSIBLE NOISE EXPOSURES [1]

Duration per day, hours	Sound level dBA slow response
8	90
6	92
4	95
3	97
2	100
1½	102
1	105
½	110
¼ or less	115

[1] When the daily noise exposure is composed of two or more periods of noise exposure of different levels, their combined effect should be considered, rather than the individual effect of each. If the sum of the following fractions: $C_1/T_1 + C_2/T_2 + \cdots C_n/T_n$ exceeds unity, then, the mixed exposure should be considered to exceed the limit value. Cn indicates the total time of exposure at a specified noise level, and Tn indicates the total time of exposure permitted at that level.

[1910.95 Table G16 amended at 39 FR 19468, June 3, 1974]

Exposure to impulsive or impact noise should not exceed 140 dB peak sound pressure level.

Figure 2.1. OSHA noise regulation.

3 - OVERVIEW OF NOISE PROBLEMS IN THE FOOD PRODUCTS INDUSTRY

Based on U.S. Department of Labor statistics, the estimated costs for noise abatement in food plants ranks in the top four among all types of American industries. In addition, the food process industry faces constraints in solving noise problems which are not imposed upon other industries. All noise control materials and systems installed in food plants must comply with strict sanitary standards. FDA and USDA requirements eliminate from consideration many of the most effective acoustical materials.

The spectrum of noise sources which are encountered in food plants is also the greatest of any industry. In the textile industry for example, the majority of noise problems are confined only to looms, twisters, and braiders. In the various types of food plants, literally hundreds of basic types of equipment, ranging from animal kill rooms to packaging, are commonly found to contribute to noise problems. This diversification of noise generating mechanisms makes any concentration of research efforts impossible. Each type of problem must be considered individually. The result is that the majority of recent research on noise control has been directed toward other areas (presses, looms, etc.).

Notwithstanding these challenges, significant advances have been achieved in recent years in many types of food process plants. This report presents most of these recent advances in the form of solution approaches for noise problems in meat packing plants, poultry plants, dairies, vegetable and fruit canneries, bakeries, nut processing plants, and beverage plants.

4 - FEASIBILITY

To establish that solution of a noise control problem is feasible, one must consider three areas:

- Acoustical Feasibility
- Production Feasibility
- Economic Feasibility

To establish acoustical feasibility, it must be shown that designs exist which would provide adequate noise reduction.

Each proposed noise control design must be reviewed to insure suitability to the application for which it is intended, and to establish production feasibility. Non-acoustical considerations related to any design include:

a. Employee safety and hygiene.

b. Fire code compliance.

c. Operational integrity:
   1. accessibility to equipment
   2. maintenance serviceability assurance
   3. product quality assurance

d. Machine system compatibility:
   1. mechanical (power, speed, etc.)
   2. service life
   3. ventilation and cooling

Figure 4.1 illustrates the matrix of decisions to be made in determining feasibility. In cases where doubt arises as to acoustical or production feasibility, a design prototype may be required.

The authors of this study, as acoustical consultants, have utilized the following design investigation procedure to establish a basis for acoustical non-feasibility for several industrial operations:

1. A literature search is performed of all available publications in the noise control field and in the general field of the alleged violation.

2. The problem is discussed with colleagues within the professional community to identify where potential solutions to the problem may have been attempted.

3. Recognized authorities in the academic community are solicited for ideas.

4. The literature of all manufacturers of acoustical materials and systems is reviewed for solution approaches, and many are contacted personally.

Feasibility

5. The manufacturer of the noise-producing equipment is contacted, as are several manufacturers of similar equipment.

6. Trade associations are contacted, and the industry-wide state-of-the-art is sought.

7. Solution approaches are solicited from the OSHA personnel involved in the citation.

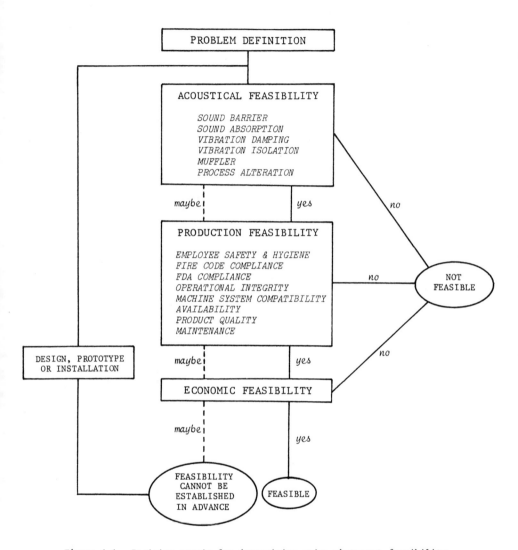

Figure 4.1. Decision matrix for determining noise abatement feasibility.

5 - LITERATURE SEARCH

A complete search of publications since 1970 revealed very few articles relating to noise control in food plants. The search included review of periodicals serving both the food industry and the noise control community. The articles which do exist are perdominantly of two types:

  a. Several articles published after the promulgation of the Occupational Safety and Health Act discuss implications of the new law, and only briefly discuss specific problems encountered by the food industry.

  b. Recent articles emphasize conflicts between the FDA/USDA guidelines and OSHA noise requirements. Again, these articles do not discuss solutions to specific problems in the industry.

The following bibliography summarizes publications relating to noise control in the food industry:

1. "Airport field or Soft Drink Plant," *Softdrinks*, June 1972, pp. 62-64.

    The article presents a warning to soft drink plants regarding impact of the new Occupational Safety and Health Act.

2. Carey, W. W., "The Ramifications of Doing Noise Control in Food Plants," *Proceedings*, Noisexpo 76, March 29, 1976, pp. 78-80.

    The conflicts facing food plants in complying with OSHA noise standards in addition to adhering to stringent sanitation guidelines of FDA, USDA, municipalities and Company Quality Assurance Laboratories and presented. Difficulties are also encountered in the use of polyurethane foam covered by a protective layer of Tedlar or Mylar film because of both potential sanitation problems and degraded sound absorptive properties. Cost considerations are also discussed.

3. Deihl, G. M., "Environmental Effects on Machine Noise in Food Processing Industries," *Compressed Air*, Vol. 81, No. 4, April 1976, pp. 6-9.

    The effect of hard wall surfaces in food plants on reverberant noise buildup is quantitatively described.

4. Draebel, J., "The Reduction of Noise in Bottle & Can Handling," *International Bottler and Packer*, December 1972, pp. 45-50.

    Design parameters of bottle and can conveying systems are assessed as they relate to noise control. Concepts covered include: modulation of containers, flow patterns for combiner tables, control of container velocity, influence of conveyor friction, transfers, and transfer plate design.

5. Elliott, R. A., "Noise Pollution and the Dairy Industry," *American Dairy Review*, August 1972, pp. 18-21.

    The influence of the Walsh-Healy Public Contracts Act, Williams-Steiger Occupational Safety and Health Act, and Chicago Noise Ordinance are discussed. Several specific noise sources of the industry are briefly discussed.

6. Klick, D. C., "Can You Hear Above the Noise?" Northwest Food Processors Association, February 1974.

    The statement is made that "Occupational noise is OSHA's greatest challenge to the food processing industry." The paper outlines what you should know about noise, what you should do about noise, and where you can get help to tackle noise problems.

7. Kugler, B. A., and Klick, D. C., "Noise Control Design Guide for the Food Processing Industry," presented at the Acoustical Society of America, 1975.

    This paper presents a brief overview of an industry-wide program sponsored by the Northwest Food Processors Association, which resulted in the development of controls for six major noise sources in the industry: (1) vibrating conveyors, (2) blanchers, (3) positive and negative air systems, (4) high pressure air and steam, (5) can handling systems, and (6) seamers.

8. Melling, T. H., "Noise in the Brewing Industry - The Sources of Its Control," Inter Noise *Proceedings*, October 1972, pp. 313-317.

    Typical sound power levels in bottling lines and kegging lines are presented. Techniques for sound reductions include screens, enclosures, control of bottle pile-ups, conveyor lubrication.

9. Melling, T. H., Wood, B. R., "Noise Generation and Prediction in Automated Bottling Lines," *Noise Control Engineering*, September/October 1974, Vol. 3, No. 2, pp. 66-70.

    Measurements are reported for a bottling plant, including washing and filling. Sound power levels were 114 dBA for the filler and 104-106 dBA for bottle-to-bottle impacts. Calculations show the effect of bottle size and velocity on noise levels.

10. Miller, R. K., "You Can Reduce Food Plant Noise," *Food Engineering*, March 1973, pp. 133-140.

    This is the first of a two part series which dealt with sources of noise, its measurement, general engineering approaches, and development of an abatement program.

11. Miller, R. K., "Reducing Noise in Food Plants," *Food Engineering*, February 1974, pp. 75-76.

    Materials for use in food plants are discussed: barium-impregnated vinyl sheet and Tedlar-faced polyurethane foam. Application of noise control engineering in food plants is mentioned: soybean flaking mill, impact noise, ceiling absorption in pecan cracking room, vibration damping on hoppers, enclosure for icecream bar wrapper machine, air noise silencers, isolation for pulverizer, isolation of dairy equipment, and bottling line modifications.

12. Miller, R. K., "Acoustical Materials for the Food Process Industry," *Proceedings* Noise Con 73, October 1973, pp. 519-522.

    Sound absorption coefficients are presented for open-cell polyurethane foams faced with Tedlar, a material applicable in some food plant situations. The use of barriers is also discussed.

13. Mitchell, N. E., "The Occupational Safety and Health Act and its Application to the Candy Industry," *The Manufacturing Confectioner*, July 1972.

    The hazard of noise is included in this general article relating to the new OSHA standard.

14. Schwartz, A., "In-Plant Noise Exposure Control in a Chocolate Factory," *Proceedings*, Inter Noise 75, pp. 215-218.

    Noise exposure levels and approaches to noise control for a chocolate factory in Israel are presented.

15. Semling, H. V., "OSHA Hears Food Processing Industries' Objections to Noise Exposure Rule," *Food Processing*, October 1975, pp. 26-32.

    This article covers participation by food processing spokespersons at the July and August 1975 hearings regarding the proposed OSHA noise regulation revisions.

16. Semling, H. V., "New OSHA Noise Proposals May Cost Industry 1.7 Billion," *Food Processing*, September 1975, p. 21.

    The potential economic impact to food plants indicated by the revised OSHA economic impact study is presented.

6 - ECONOMIC ANALYSIS

An economic impact study prepared by Bolt, Beranek and Newman and released by the U.S. Department of Labor on 17 June 1976 estimated the following for food product manufacturers (SIC 20):

Compliance cost, 90 dBA:	$575,000,000
Compliance cost, 85 dBA:	$1,675,000,000
Monitoring costs (annual):	$13,500,000
Audiometric costs, 85 dBA (annual):	$6,300,000

This estimate included the following breakdown by specific industry (85 dBA):

Industry Type	Cost, Millions of Dollars
Meat	$536.2
Canned and Preserved Fruits and Vegetables	206.2
Fats and Oils	161.7
Beverages	17.1
Sugar and Confectionary Products	16.1
Bakery Products	12.1
Dairy Products	9.3
Miscellaneous Products	25.1

It was also estimated that 315,400 employees of a total 1,126,300 production workers in the food industry are exposed to sound levels exceeding 85 dBA; 180,200 to more than 90 dBA.

The publication of this economic impact study met with considerable industry-wide criticism, particularly from the food industry. To date more accurate cost estimates have not been recognized, however. Due to the controversies which have arisen, it is suggested that the above costs be considered only as an "order of magnitude" estimate, which also possibly may be used to assess the relative severity of noise problems in various types of food plants.

7 - GOVERNMENTAL SANITARY REQUIREMENTS

Operating under the Federal Food, Drug and Cosmetic Act, the Food and Drug Administration (FDA) is responsible for regulating all foods except red meats, poultry and eggs, which are the responsibility of the Department of Agriculture (USDA).

Information was recently obtained from the FDA and USDA regarding requirements for noise control materials. Copies of these letters follow this chapter.

In general, all materials in product contact zones should comply with the following requirements:

I. *Non-Toxic*

The USDA utilizes FDA approval lists and procedures for approval of non-toxic materials. A list of approved materials is provided in the Food Additive Regulation, Code of Rederal Regulations, Section 121.25. Approval may be obtained for other materials by providing (1) trade or brand name, (2) use, and (3) chemical composition to:

> Mr. John W. Sloan
> Scientific Services
> Animal & Plant Health Inspection
>   Service
> U.S. Department of Agriculture
> Washington, D.C. 20250

II. *Impervious*

The material must be impervious to water, meat juices, etc.

Our investigation has indicated that the most compatible materials for noise control would be stainless steel and certain plastics such as Teflon, PVC, etc. It may be noted that USDA approval has been granted to a PVC curtain impregnated with a metallic lead; however, lead salts are not allowable.

Regarding materials not in food contact, the primary requirement is the prevention of porous or irregular surfaces which permit dirt or bacterial build-up. The material must be cleanable. MPI Bulletin 350 directs enforcement against open polyurethane foams for walls and ceilings. Foams with an impervious facing, however, have received USDA approval.

A discussion of sanitation considerations related to noise control in food plants was presented by Dr. Carey at NOISEXPO 76. A copy of this presentation follows this chapter.

DEPARTMENT OF HEALTH, EDUCATION, AND WELFARE
PUBLIC HEALTH SERVICE
FOOD AND DRUG ADMINISTRATION
WASHINGTON, D.C. 20204

January 31, 1977

Mr. Richard K. Miller, President
Richard K. Miller and Associates, Inc.
464 Armour Circle, N.E.
Atlanta, GA 30324

Dear Mr. Miller:

This replies to your letter of January 4 concerning noise control materials for use in food processing plants.

The Federal Food, Drug, and Cosmetic Act contains no specific requirements concerning building construction or building equipment used in food processing plants. However, the various adulteration provisions of the Act must be taken into account in considering the construction and types of building fixtures and equipment used in a plant intended for processing.

The Act provides, among other things, that food is adulterated if it has been prepared, packed, or held under insanitary conditions whereby it may have become contaminated with filth, or whereby it may have been rendered injurious to health. We would expect that materials used in building construction to be of a nature so that the food processed in its vicinity would be adequately protected from contamination with filth or other substances deleterious to health. The construction employing the material should be of a design to facilitate inspection and cleaning and should be rodent proof and should not provide harborage for insects or other vermin.

Sincerely yours,

Pat T. Adamo
Assistant to the Director
Division of Regulatory Guidance
Bureau of Foods

UNITED STATES DEPARTMENT OF AGRICULTURE
ANIMAL AND PLANT HEALTH INSPECTION SERVICE
WASHINGTON, D.C. 20250

March 25, 1977

Mr. Richard K. Miller
President, Richard K. Miller and
  Associates, Inc.
464 Armour Circle, NE.
Atlanta, GA 30324

Dear Mr. Miller:

This is in response to your recent request for information as to what
materials intended for noise control can be used in food processing plants.

Our jurisdiction covers those meat and poultry plants that operate under
the Federal inspection program. We require materials which represent a
surface in any room where meat or poultry products are prepared, handled,
or stored to be chemically acceptable and possess the physical characteristics associated with cleanability. Basically, that implies smoothness,
firmness, and imperviousness. The material should not have cracks or crevices
which make cleaning very difficult, and should be firm enough to resist
the normal abuse that will occur whether it occurs from operation activity
or cleanup. Obviously, the use technique of the material can affect the
acceptability. In other words, a material may be acceptable both from
a chemical and physical standpoint; but, the construction design used in
effecting noise control may result in the finish item not being acceptable.
Any noise control design cannot cause unsanitary conditions to occur.

Requirements in other food producing plants would fall under Food and
Drug Administration's jurisdiction and any local codes that may exist.

Sincerely,

K. E. Peterson
Senior Staff Officer
Facilities Group
Technical Services
Meat and Poultry Inspection Program

-15-

UNITED STATES DEPARTMENT OF AGRICULTURE
Animal and Plant Health Inspection Service
Meat and Poultry Inspection Program
Washington, D.C. 20250

MPI BULLETIN 350

ACTION BY: Regional Directors, Area and Circuit Supervisors

INFORMATION FOR: Plant Management and State Officials

Polyurethane Foam

A limited number of installations have used "foamed-on" polyurethane for walls and ceilings and, in some cases, to insulate drip pans. Conflicting reports have been received on its adequacy as a sanitary surface.

From observations and discussions with field personnel, we are now convinced that polyurethane foam does not make an acceptable surface. It cannot be consistently applied so that it is smooth, hard, impervious, and capable of being cleaned.

Effective immediately, please do not approve installations of polyurethane unless covered by a suitable smooth, washable surface--cement plaster, tile, approved panels, etc.

Existing installations will be judged on their merits. Corrections should be required if the polyurethane-covered surface cannot be kept clean, or if it may contaminate product. No action is needed if the surfaces are capable of being readily and thoroughly cleaned.

Fred J. Fullerton
Deputy Administrator
Field Operations

DISTRIBUTION: Q, P, U-U-2

OPI.STS:PFE
July 17, 1973

● U. S. GOVERNMENT PRINTING OFFICE : 1973 841-800/2

from

"The Ramifications of Doing Noise Control in Food Plants"
by Dr. Walter W. Carey
(presented at Noisexpo 76)

A complete presentation of the conflict between sanitation and noise exposure requirements should take into account a comparison of the requirements of the FDA, USDA, state, and municipal regulations with those pertaining to employee noise exposure. Since the present intent may be served by reference to only one such set of standards, for the purpose of this presentation, the discussion shall be limited to a comparison between FDA and OSHA requirements. USDA and other regulations are similar in intent to those of the FDA and compliance with the intent of the latter would certainly be required in any actual case. A good discussion of the problems faced by the dairy industry in complying with the USPHS "Grade A pasteurized Milk Ordinance" and OSHA standards has been prepared by Greiner[2] and reference is made to that document for further details.

Present FDA regulations dealing with GMP's are contained in Title 21, Code of Federal Regulations, Part 128.[3] Part 128 is generally recognized as an "umbrella" GMP regulation applying to all facilities engaged in the manufacture or processing of foods for human consumption and sets forth requirements for sanitation and cleanliness that must be complied with to assure that such food is safe and has been prepared, packed, and held under sanitary conditions.

Section 128.3 (b)(1)--Plant Construction and Design--specifically relates to conditions within the actual facility and, in part, requires that "... Floors, walls, and ceilings in the plant shall be of such construction as to be adequately cleanable and shall be kept clean and in good repair". The key word here is "adequate" which is defined in Section 128.1 (a) to mean "...that which is needed to accomplish the intended purpose in keeping with good public health practice". What is being said here, in effect, is that it is up to the manufacturer, and not the agency, to determine the means whereby the workplace may be maintained in a condition of sanitation and cleanliness acceptable to and consistent with standards of good health and the intent of the regulation. In other words, the burden of development, proof, and compliance is placed squarely on the manufacturer which is quite removed from the regulations of certain of the other agencies, notably OSHA and EPA, which in many instances spell out in minute detail what must be installed and how it is to be operated to assure compliance with the promulgated standards.

A further indication of the requirements imposed by these "umbrella" regulations is contained in a subsequent passage in Section 128.3 (b)(1) which states "...Fixtures, ducts, and pipes shall not be so suspended over working areas that drip or condensate may contaminate foods, raw materials, or food contact surfaces". This requirement, taken in conjunction with that discussed previously, in effect limits the application of acoustical absorption materials in the workplace to reduce reverberant build-up due to hard floor, wall, and ceiling requirements. Installation of acoustical absorption panels is, in many cases, limited by this regulation to wall areas so as not to provide surfaces for contaminant build-

up over product lines or equipment. These installations may, and usually do, result in decreased area which is available for absorption as well as difficulties in ensuring that the space between the panel and wall surfaces is sealed against contamination and/or infestation or designed so that access may be had on a routine basis for cleaning.

Additional references limiting the use of acoustical materials in the food environment are found in Section 128.4(a) which states that "...All equipment should be so installed and maintained as to facilitate the cleaning of the equipment and of all adjacent spaces"; Section 128.6(c) goes on to state "...All utensils and product-contact surfaces of equipment shall be cleaned as frequently as necessary to prevent contamination of food and food products. Nonproduct-contact surfaces of equipment used in the operation of food plants should be cleaned as frequently as necessary to minimize accumulation of dust, dirt, food particles, and other debris". These sections, taken together, place severe restriction on the use of acoustical enclosures around noisy equipment since the enclosures must be both tight fitting to prevent acoustical leakage and infestation and, at the same time, easily removable to allow for cleaning and maintenance operations. Additionally, all crevices must be minimized (i.e., those for access panels and breakdown joints), since these are prime locations for accumulation of dust or otherwise.

The above excerpts from the so-called "umbrella" GMP apply to and set forth standards for the entire food industry; however, in addition to these requirements, FDA has also seen fit to develop specific regulations for the various segments of the industry in order to provide more specific guidelines on plant sanitation practices. To date, four such separate standards have been developed: Part 128(a)--Fish and Seafood Products, Part 128(b)--Thermally Processed Low-Acid Foods Packaged in Hermetically Sealed Containers, Part 128(c)--Cocoa Products and Confectionary, and Part 128(d)--Processing and Bottling of Bottled Drinking Water.[3] Each regulation establishes specific requirements, for the affected industry segment, which serve to strengthen those of the "umbrella" GMP and further restrict usage of common acoustical materials.

Additional citation to the GMP's could be made; however, the case has, I believe, been made that unrestricted use of acoustical materials and standard acoustical treatment cannot be tolerated in the food industry. To devote all efforts at compliance with one or the other agency's regulations will inevitably result in regulatory enforcement by the agency whose standards have not been adhered to. Penalties for violation of OSHA standards are well-known; however, the FDA can also take enforcement measures which include confiscation and destruction of articles deemed to have been manufactured under noncompliance conditions. Additionally, FDA compliance inspections can and do result in notification of violations of the pertinent standards and such violations are issued, to the facility under investifation, for abatement.

-17-

8 - SANITARY DESIGN PRINCIPLES FOR NOISE CONTROL INSTALLATIONS

The following guidelines are applicable to the installation of noise control systems (adapted from "Sanitary Design Principles" by William S. Stinson, *Food Processing*, October 1976).

1. Avoid locating mechanisms directly above the product stream.

2. Cover the product stream.

3. Bearings and seals should be located outside the product zone. Parts in product zone must be sealed or self lubricated.

4. Minimize loose items.

5. All filters must be easily replaced and cleaned.

6. Systems should be designed for thorough washdown with high-temperature (140-180°F) water and detergents as required for proper cleaning. In some plants high pressure (600 psi) water is used.

7. Building contractors should be aware of necessity of keeping site clear to prevent attracting rodents and insects.

8. False walls and voids in walls should be avoided, particularly in process areas.

9. Avoid using glass in, above, or near process areas.

10. Select or design equipment to be safe under processing conditions, easily cleaned and inspected.

11. Construction materials should be selected to resist wear and corrosion and to protect contents from external contamination.

12. Product contact surfaces must be inert, non-toxic, non-porous, smooth (no cracks, crevices, or sharp corners), easily cleaned, non-peeling, and inert to steam cleaning, hot water, and sanitizing solutions.

13. All inside corners should have internal angles of sufficient radius (1/4" or greater) to provide easy cleanability.

14. All surfaces in contact with food should be visible for inspection or readily disassembled for inspection. It should be demonstrated that routine cleaning procedures eliminate contamination from bacteria, insects, and soil.

15. Painting of product contact and product zone surfaces should be avoided.

16. All surfaces in contact with food should be readily accessible for manual cleaning. If not readily accessible, equipment should be easily disassembled for manual cleaning. If mechanical in-place cleaning is used, the results achieved without disassembly should be equivalent to those obtained with manual cleaning.

17. Interior surfaces in contact with food should be self-emptying or self-draining.

18. Exterior of equipment must be easily cleaned and not retain soil or wash water.

19. If equipment requires adjustment during operation, it should be designed so operators do not place their hands in product zone.

20. Floor attachments should be minimized.

21. Equipment should be at least 8" off the floor, mounted on single pedestals wherever possible, and with clearance of at least 18" from ceiling. Use single point contact supports.

22. Equipment should be mounted at least 36" from a wall and at least 36" should be allowed between equipment.

23. All openings into equipment should be protected against entrance of contaminants as a function of the action of opening or the open condition.

24. Internal pipe caulking should be avoided.

25. Flexible piping should be non-porous, not affected by the food or cleaning compounds and in sections not over 3' long.

26. Piping must be designed to operate "flooded" during normal operation.

27. Avoid using screwed pipe in product lines Welded pipe must be inspected for proper penetration of weld before installation.

28. Select gasket materials to provide proper seals and not contaminate the product.

29. Continuously weld all support connections.

30. Seal all ends of support members.

31. Floors, walls and ceilings must be smooth, non-peeling, inert to process and easily cleaned.

32. Floor, wall and ceiling corners should be coved for easy cleaning. Suggest 4" radius.

33. Caulk or seal all wall, floor, and ceiling joints.

34. Structural members must be integral to the supported surface or caulked to it.

35. Avoid painted walls and ceilings, particularly where moisture is involved. Use prefinished, easily cleaned panels, insulated as necessary.

9 - MATERIALS INVENTORY

A survey was conducted in 1973, and updated in 1977, inventorying over 200 manufacturers of acoustical materials with products which are applicable to the food industry. The results of this survey are presented in Table 9.1. Where FDA or USDA approval is indicated, it should be recognized that the material(s) involved are approved as food contact surfaces; however, each application of these materials must be reviewed in terms of each specific system or installation. Additional information on some of these products is presented in Chapter 31.

TABLE 9.1

MANUFACTURERS OF ACOUSTICAL PRODUCTS FOR THE FOOD PROCESS INDUSTRY

MANUFACTURER	USDA APPROVED	FDA APPROVED	STEAM CLEANABLE	STAINLESS STEEL	BARRIERS	FACED ABSORBERS	MUFFLERS	PIPE COUPLINGS	ENCLOSURES	DAMPING	DOORS
Allied Witan Company 12500 Bellaire Road Cleveland, OH 44135							●				
American Acoustical Products 9 Cochituate Street Natick, MA 01760			●	●	●	●					
Body Guard, Inc. P.O. Box 8338 Columbus, OH 43201		●			●	●			●		
Doug Biron Associates, Inc. P.O. Box 413 Buford, GA 30518				●	●	●	●	●	●	●	●
Dow Corning Midland, MI 48640		●								●	
Eckel Industries, Inc. 155 Faucett Street Cambridge, MA 02138		●			●	●					●
Ferro Corporation 34 Smith Street Norwalk, CT 06852	●				●	●			●		
Fluid Kinetics Corporation P.O. Box C.E. Ventura, CA 93001			●				●				
B. F. Goodrich P.O. Box 657 Marietta, OH 45750	●	●			●						●
I.D.E. Processes Corporation 106 Eighty-First Avenue Kew Gardens, NY 11415			●			●	●		●		
Industrial Acoustics Company 1160 Commerce Avenue Bronx, NY 10462									●		

TABLE 9.1 (Continued)

MANUFACTURER	USDA APPROVED	FDA APPROVED	STEAM CLEANABLE	STAINLESS STEEL	BARRIERS	FACED ABSORBERS	MUFFLERS	PIPE COUPLINGS	ENCLOSURES	DAMPING	DOORS
Industrial Noise Control 785 Industrial Drive Elmhurst, IL 60126				●	●	●				●	
Korfund Dynamics Corporation Cantiaque Road Westbury, NY 11590	●	●		●	●						
Metal Building Interior Company 1176 E. 38th Street Cleveland, OH 44114						●	●				
Mercer Rubber Company 136 Mercer Street Trenton, NJ 08690	●							●			
Martec Associates, Inc. 1645 Oakton Street Des Plaines, IL 60018					●		●				
Noise Control Products, Inc. 969 Lakeville Road New Hyde Park, NY 11040					●	●	●	●			
Noise Measurement and Control Corp. 322 E. Lancaster Avenue Wayne, PA 19087					●	●	●				
Rollform, Inc. P.O. Box 1065 Ann Arbor, MI 48106							●				
Scott Foam Division 1500 E. Second Street Chester, PA 19013		●				●					
Soundcoat, Inc. 175 Pearl Street Brooklyn, NY 11201						●				●	
Specialty Composites Delaware Industrial Park Newark, DE 19711	●			●	●					●	

TABLE 9.1 (Continued)

MANUFACTURER	USDA APPROVED	FDA APPROVED	STEAM CLEANABLE	STAINLESS STEEL	BARRIERS	FACED ABSORBERS	MUFFLERS	PIPE COUPLINGS	ENCLOSURES	DAMPING	DOORS
Titeflex 603 Hendee Street Springfield, MA 01109	●							●			
Vanec 2655 Villa Creek Drive Dallas, TX 75234			●					●			
Veneered Metals, Inc. P.O. Box 327 Edison, NJ 08817	●		●	●						●	

10 - MEAT PACKING PLANTS

Acoustical studies of three meat packing plants identified the following major noise producing operations:

    1. Splitting saws
    2. Choppers
    3. Hydroflakers
    4. Emulsitators
    5. Ham curing machines
    6. Weiner and sausage peelers
    7. Packers and vacuum machines

In addition to these items of equipment, other types of machinery may generate excessive sound levels due to motor noise (see Chapter 23), air exhaust (see Chapter 24) and metal-to-metal impacts (see Chapter 25).

The following guidelines may be followed in developing noise control techniques for the six major items of equipment.

SPLITTING SAWS

Sound levels of a splitting saw are typically 100-105 dBA. Since this noise is intermittent, it is important that actual exposure time be carefully assessed. One example is of a splitting saw operation where sound levels averaging 100 dBA are generated during 20 second cuts for each of 540 animals processed per shift. The daily exposure time is three hours (540 times 20 seconds). OSHA allows a daily exposure of 2 hours to a sound level of 100 dBA (Figure 2.1). Thus the dosage is:

$$D = \frac{3}{2} = 150\%$$

The redesign of existing saws is quite an impractical task for the user. It does not appear feasible to reduce the noise generated by existing splitting saws by means of redesign. However, new, quieter band-type saws are available from Jarvis and Bess & Donovan.

One method to reduce employee noise exposure to achieve OSHA compliance is to rotate employees to reduce the daily exposure time. In the above example, the rotation of two workers would result in an exposure of 75% for each, assuming the remainder of the work day was spend in an area with sound levels below 90 dBA. In some plants splitting saw operators are already rotated due to the strenuous nature of this task. In other plants, rotation may not be feasible due to skills required, union stipulations, safety, etc.

The noise generated by the splitting saw may be isolated from adjacent employees by the installation of a barrier. This barrier may be a conventional wall with minimum weight of 1.0 psf.

CHOPPERS

The sound levels of choppers range from 100 to 105 dBA. Chopper noise is primarily due to blade vibration. Our previous studies have indicated that modifications of the blade or rotational speed variation (within feasible limits) will not satisfactorily reduce chopper noise.

To the best of our knowledge, successful designs have not been implemented to effectively reduce the noise levels of choppers. If engineering development is undertaken, however, it may be feasible to develop a practical design solution to this problem. The following approaches may be considered:

1. An airtight, hinged, gasketed and latched cover may be installed on the chopper. The cover should be fabricated from austentitic stainless steel, such an AISI type 304 or 308. It should be constructed of guage 7 U.S. standard (.1875") or thicker sheet. Gasketing may be accomplished through the use of a thin-walled tubular-shaped gasket. Meat may be fed into the chopper by either of the following methods:

   a. Meat conveyed to chopper in removeable bottom tub by personnel.

   b. Tub lifted to feed hole with a hoist.

   c. Feed hole uncovered and meat fed into chopper by gravity.

   d. Feed hole resealed during chopping operation.

   e. A screw conveyor.

2. As an alternative, a sliding door may be provided in the cover. This door must seal airtight during the chopping operation. A rotary deflector which moves vertically could be used to revove meat from the chopper.

3. The blade area of the chopper may be enclosed with stainless steel or USDA approved non-metallic flaps which extend from the existing cover to the meat surface.

HYDROFLAKER

The sound levels of one hydroflaker ranged from 100 to 105 dBA. The primary noise generating mechanism was vibrations induced by the drive mechanism into four panels:

   a. The top surface
   b. The machine back
   c. The back, hinged cover
   d. A large section of the conveyor

Since redesign of the drive mechanism was not considered practical in this case, it was recommended that a constrained layer damping treatment be applied to these panels. This treatment consisted of a viscoelastic layer, such as:

Product: EAR C1002
Available from: Doug Biron Associates, Inc.
P.O. Box 413
Buford, Georgia 30518

Product: Dyad
Available from: Soundcoat Company
175 Pearl Street
Brooklyn, NY 11201

faced with a layer of 24 GA stainless steel.

EMULSITATORS

The sound levels of emulsitators range from 94-98 dBA. The primary noise is due to vibration of the outlet piping. To reduce this noise, a flexible coupling 2" to 5" in length, constructed of an FDA-approved plastic such as PVC, reinforced with stainless steel mesh, may be installed at the pump outlet prior to the piping. This may necessitate additional support of the pipe. Additional noise reduction may be achieved by treating the exterior surface of emulsitator hoppers with a constrained layer damping treatment such as discussed above.

HAM CURING MACHINES

Sound levels of 96-99 dBA are typical of ham curing or ham injecto machines. This noise is identified as being due to pump outlet piping vibration, and may be reduced by installing a flexible coupling on each pump outlet. The coupling should be of an FDA-approved plastic, reinforced with steel mesh, and specified to the pump outlet pressure.

As an alternative to this isolation, the pump section may be enclosed by use of covers and sealing the enclosure airtight. In addition, the interior of the enclosure may be lined with a 1" plastic-faced acoustical foam, and the piping extending outside the enclosure may be wrapped with an acoustical lagging treatment. Ventilation of the enclosure should be provided to prevent motor overheating.

WIENER AND SAUSAGE PEELERS

Wiener peeler noise is primarily due to the air flow which separates the casing from the wiener. The following approaches to noise control may be considered:

1. Air-steam shut-off switch: The addition of a control switch to stop

the flow of air when wieners are not being peeled will reduce the operator's exposure to noise. Such a switch may be controlled by one of two methods:

a. A foot switch may be installed to allow the operator to start and stop the air flow. This method of control is dependent on the operator and is therefore subject to human error.

b. A switch at the entrance to the steam box, sensitive to the presence of wieners may be installed. Such a switch can be either light sensitive or pressure sensitive. This method of control will work independent of the operator and is therefore the more desirable.

It should be mentioned here that the implementation of this system will result in lower air costs. If steam is also controlled in this fashion (by the same switch, in phase with the air), there will be a double savings.

2. Air nozzle modification: The major source of noise of the wiener peeler is the air jet used to separate the split skins. Air is ejected through three 1/16" diameter ports. The ports may be increased to 1/4" diameter, and an AE Collimator Insert muffler should then be fitted into each hole. These units are available from:

> Allied Witan Company
> 12500 Bellaire Road
> Cleveland, OH 44135

Our calculations show that similar air thrust force and slightly lower air consumption will result from use of such a muffler. This approach should be tried on an experimental basis. We are not aware of its ever having been employed on production peeler units. Additional approaches to noise control involving air jets utilized for their thrust characteristics are presented in Chapter 15.

3. Acoustical enclosure: A partial or total enclosure of the peeler may be installed. The primary noise path is from the output side of the machine. In some cases partial enclosure design may be incorporated utilizing the peeler as a frame.

4. Air flow regulation: Experiments should be conducted to determine the air and steam requirements for each product. Flow should then be regulated (independent of the operator) at the air (or steam) line. This action will reduce noise levels and air-steam costs.

PACKERS AND VACUUM MACHINES

These machines are generally pneumatically operated and frequently have unsilenced air discharges. Appropriate mufflers should be installed (see Chapter 24).

11 - POULTRY PLANTS

Sound level surveys in poultry plants may indicate excessive sound levels due to three operations:

- Picking
- Lung Guns
- Packing

The following approaches to noise control should be considered:

PICKING

Noise in picking areas is generated by picker brushes, gears, and heating flames. Continuous attention by operators is generally not required, and noise exposure times may be quite low. The picking noise, however, may carry over into adjacent areas, exposing other employees. This noise may be most easily controlled by confining it to the picking room. Specifically, this may be achieved by the construction of a complete wall, such as of painted concrete block, isolating the picking area. Product should be passed through an opening in the wall, which must be of minimum area to reduce sound transmission. The opening may also be baffled.

LUNG GUNS

Suction noise is a dominant source in lung guns. This noise may potentially be reduced by providing a complete seal around the opening in the bird to confine the noise. Air is then supplied to the interior volume by means of a separate air hose.

It was reported that such a product was being developed by Gamesco of Gainesville, Georgia.

PACKING

Noise in packing areas is often generated by vibrations of the stainless steel work tables due to impact of product and other objects. This noise may be reduced by treating the underneath side of the tables with a vibration damping treatment. Due to its remote location, the damping treatment should not prevent sanitation problems; however, precautions should be observed to select a material which is smooth, cleanable and non-toxic (see Chapter 8).

12 - DAIRIES

The acoustical environment of dairies consists of the noise of numerous items of mechanical equipment which is amplified by the reverberation of the work space due to hard wall, ceiling and floor surfaces. In some cases, reverberant noise levels may be reduced by sound absorptive treatment (see Chapter 27). Generally, however, noise reduction must be achieved by means of equipment modifications. Approaches to equipment noise control are presented in the following sections.

FILLERS

Gas fired burners used on some fillers generate sound levels in excess of 90 dBA. Ex-Cell-O Corporation has developed two quiet burner models which reduce the burner noise to 76 dBA at 3 feet. The cost of replacement burners is approximately $700. It should be pointed out that some fillers use electric heaters rather than burners, eliminating this noise source.

Another source of noise is the 92 to 94 dBA generated by using compressed air to discharge the cartons from the bottom forming section through a chute to the rear conveyor chains. A prototype machine developed by Ex-Cell-O is presently running in the field, which uses a mechanical stripper device, eliminating the air in this location. As a result, a reduction of noise was accomplished as a reading of 82 dBA was recorded at a distance of 3 feet while the machine was running. The cost of the modification is approximately $1000.

Other noise sources on fillers include motors (see Chapter 23), air discharges (see Chapter 24), filler bowl vacuum pumps (see Cahpter 19), and vibrators (see Chapter 16).

MATERIALS HANDLING - AUTOMATIC CASERS AND STACKERS

The primary noise sources of these items of equipment are air exhaust (see Chapter 24) and metal-to-metal impacts (see Chapter 25).

HOMOGENIZERS

One dairy reported a noise reduction from 107 to 97 dBA by boring the head and replacing the motor. Other potential machine design changes may include the use of a sound absorptive material within the hood, mufflers, and vibration damping. These modifications could most easily be accomplished by the machine manufacturer, and may not be practical for existing equipment.

The best approach for noise abatement of homogenizers already installed is the use of acoustical enclosures or to isolate the equipment in acoustical rooms.

SEPARATORS AND CLARIFIERS

The method of standardization and clarification of milk presents the problem of the separator/clarifier location. In many new installations the separator/clarifier is an integral part of a high-capacity HTST system. Unfortunately, the noise level of most separator/clarifiers does not meet the minimum allowable noise standards. To circumvent this problem, the new plant layout should isolate the separator/clarifier in a separate room that is near the short-time equipment, near the main control panel, and adjacent to the processing room. Control of the machine is by the main panel operator on the processing room floor.

ICE CREAM AND POPCYCLE MACHINES

Sound levels ranging from 87-92 dBA were measured for two Vitaline machines. The major noise sources were identified to be: metal-to-metal impacts (see Chapter 25), exhaust from air cylinders (see Chapter 24) and a vibratory bowl feeder (see Chapter 20).

13 - VEGETABLE, FRUIT AND PREPARED FOOD CANNERS

The primary noise problem in most high speed canning plants is can-to-can impacts on conveyor lines. Where cans are partially filled, sound levels are a function of both the product in the can and the percentage it is filled, as shown in Figure 13.1. An in-depth study of alternate solutions to can conveyor noise control is presented in Chapter 17.

Noise control for other food canning operations is discussed in the following sections.

CITRUS PLANT

The following sound levels were measured in a citrus process plant:

Operation	*Sound Level, dBA*
Evaporators (7, outdoors)	95
Grading Room	91
Bagging	95-98
Holding Tanks	91
Compressors	95-102
Extractor Room	91-94
Blending	87
Concentrate Evaporator	98
Concentrate Grader	92
Labeling Machine	90-92
Packer	92
Cooler	92
6 Ounce Filler	95
16 Ounce Filler, Capper	94
404 Filler	97
Squeeze Presses	93

CONTINUOUS COOKERS AND RETORTS

Sound levels ranging from 94-100 dBA may be generated by continuous cookers and retort equipment. This noise may be silenced by the installation of a silencer, such as available from the equipment manufacturer, or a steam exhaust muffler, such as SM Type, available from Allied Witan Company, 12500 Bellaire Raod, Cleveland, Ohio 44135. This muffler will withstand temperatures up to 1000°F. The ideal approach to retort equipment silencing seems to be to pipe away the steam. This approach is not an acceptable method, however, since regulations require that the steam discharge be visable.

FORMING MACHINES

The sound levels of a spaghetti produce forming machine were reduced to below 90 dBA by the installation of pneumatic mufflers (see Chapter 24).

Noise generated by meat ball forming machines was identified to be due to metal-to-metal impacts of machine mechanisms. For noise abatement, the use

Vegetable, Fruit and Prepared Food Canners

of non-metallic parts should be considered, as discussed in Chapter 25.

VIBRATORY SHAKE TABLES AND CONVEYORS

Many vibratory shake tables are found to have noise levels below 90 dBA, especially where soft food products are involved. Where excessive noise is observed, it may be due to:

a. Operation above the recommended design capacity.

b. Loose parts.

c. An imbalance of the conveyor; the system must be "tuned" for optimum operation and minimum noise.

d. Motor noise.

Generally, a good maintenance program is the best approach to noise control. The use of motor silencers may also be considered (see Chapter 23). Where conveyors are located overhead, a clear plastic barrier may be installed below them to shield employees from the noise.

CUTTERS

Sound levels of vegetable cutters is generally due to the drive mechanism. Sound levels of older corn cutters were measured to be 94 dBA, while new units were below 90 dBA. Noise abatement may be achieved by gear damping, non-metallic gears, or enclosure of the drive mechanism.

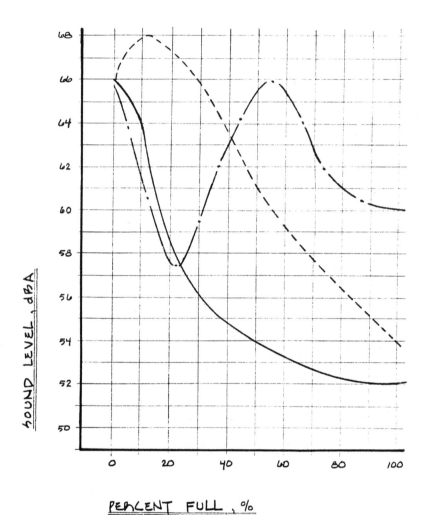

Figure 13.1. Sound levels of impacted can partially filled with water (— · —), small rocks (-------), and dirt (————).

14 - PEANUT AND PECAN PROCESSING

Noise abatement for peanut and pecan plants is considered together since they involve the direct processing of nuts and have similar noise sources. The following approaches to solving noise problems for these operations are based on experiences with four plants.

PEANUT CLEANERS

In one noise study involving cleaners by two manufacturers, it was found that cleaners in good operating condition and in proper maintenance radiated only 85 dBA; however, easily traceable maintenance problems on other machines generated noise levels in excess of 90 dBA, and up to 104 dBA. Thus, maintenance should be recognized as a basic approach to noise control.

Each cleaner should be periodically inspected in operation while all other machines in the room are off. The inspection should include measurement of the noise level of the machine at 3' on all sides. If the noise level exceeds 88 dBA, the machine elements generating the excessive noise levels should receive the required maintenance. See also Chapter 28 of this report.

The induced and forced draft blowers of cleaners are the primary noise sources. There are two principle sources of noise in a blower. The source responsible for the generation of the tone is the pulse created each time a blade passes the cut-off point. The frequency of the tone is equal to the number of times each second a blade passes the cut-off. The second major noise source in a blower is the turbulence created by the blower while adding energy to the air stream. High frequency turbulent noise is typically produced close to the blade tips while low frequency noise is usually produced in the fan inlet and outlet ducts.

There are three primary paths by which noise leaves a blower. The first of these is the blower inlet. Noise travels from its source inside the shell out through the inlet opening and then through the air to the observer. Noise also leaves the blower by way of the discharge passage, with some of this noise radiating through the walls of the blower discharge duct and toward the observer. The third noise radiating area is the housing itself. The noise created inside the housing excites the housing causing it to vibrate and the vibrating of the housing causes sound to be radiated into the room.

In order to adequately solve a blower noise problem, it is necessary to consider all these areas and treat each in accordance with the amount of sound power radiated by it. Four approaches to noise control should be considered:

        a. Inlet or outlet silencers.
        b. Lagging of the ducts.
        c. Vibration damping of the housing or ducts.
        d. Decoupling ducts.

Peanut and Pecan Processing

PEANUT SHELLERS

Sound levels of 96 dBA are typical for shellers. The primary noise sources
in shellers are the inlet noise of the husking fan and the outlet noise of
the lower blower. Noise control by use of blower silencers would be the
"ideal" solution; however, the machine structure and space constraints do
not allow for this approach. Another consideration would be a tuned, reac-
tive silencer; however, dust, complexity, and cost factors make this approach
unsatisfactory

Another approach to blower noise control is the use of sound absorptive ple-
nums for both inlet and outlet air flows. The existing machine hood presently served as a small volume plenum for both blowers. Noise control may be
achieved by lining the hood with an open cell polyurethane foam, as shown in
Figure 14.1. The foam should be four inches thick (due to the low frequency
noise spectrum of the blowers). A facing of 1 mil Tedlar should be applied
to the outer surface of the foam. The side opening of the hood should be
closed with a 1/4" thick hinged window.

It may also be necessary to isolate the outlet duct from the husking fan with
a compliant coupling, or to apply a damping treatment or lagging to the duct.

Maintenance of the shellers is also important to achieve minimum noise. In
particular, loose rods and metal parts in the vicinity of the shaking tables
may be found to generate significant noise.

PECAN GRADERS

For effective noise control of an inshell grader, the entire outer surface
area of the hoppers and chutes may be treated with a damping compound or
sheet damping, applied to a thickness of not less than 1½ times the thickness
of the metal. The easiest damping treatment technique is the application of
adhesive-backed damping sheets, applied directly to the metal surfaces after
cleaning. These materials can easily be painted for cleaning purposes. As
an alternative to the use of sheet damping, a spray-on or trowel-on damping
treatment may be used. Specific application instructions may be obtained
from these manufacturers.

PECAN POP REMOVER, WASHERS, CONVEYORS

The noise generated by this equipment can nearly always be traced to the impact of pecans on sheet metal panels. For noise abatement, the following
guidelines should be observed:

1. The back side of the panel impacted by the pecans should be treated with
   a constrained layer damping treatment.

2. The front side of the impacted panel may be treated with a wear retard-
   ant rubber material (see Chapter 25) if USDA approval is obtained.

3. Machine maintenance should be carried out to insure minimum noise levels.

nificant noise reduction. The noise shields should be constructed of
barium-loaded vinyl, 1/2 pound per square foot minimum weight with 1"
minimum thickness open-cell polyurethane foam laminated to the interior.
The face of the foam should be faced to prevent dust accumulation.

A window of Plexiglas may be incorporated in the shield for visibility
of the cracking operation. The side edges of each adjacent noise
shield should not be more than 1/8" apart, to minimize sound leakage.
The noise shields should extend to the frame supporting the machines.

A significant portion of the impact-generated noise is radiated from the
bottom of the machine, which is open. This noise path may be blocked
with a plate of 1/2" minimum thickness plywood panel faced on the machine
side with a 2" thick layer of duct liner faced glass fiber. The plywood
panel should be positioned on the angle iron frame underneath each cracker. As an alternative to individual panels, a single bottom panel may
be used on each row of crackers.

The back of the crackers should be closed with 1/2" minimum thickness
wood panels lined with 2" thick faced glass fiber on the machine side,
and should extend from the top of the hoppers to the frame. A door may
be placed at each machine for visibility and belt replacement. The panels should be constructed for each removal, such as by being fastened
with hooks and equipped with handles.

4. A layer of nylon may be positioned between the metal surfaces associated
with the machine impacts. Experiments with one cracker indicated a 2-3
dB reduction with this modification. While nylon does not provide optimum impact isolation, it seems to be the only material available which
would withstand the wear for any length of time.

5. The exterior surfaces of the pecan hoppers may be treated with a vibration damping material.

6. A sound absorptive material may be installed in the cracking room if
reverberant sound build-up is identified as a problem (see Chapter 22).

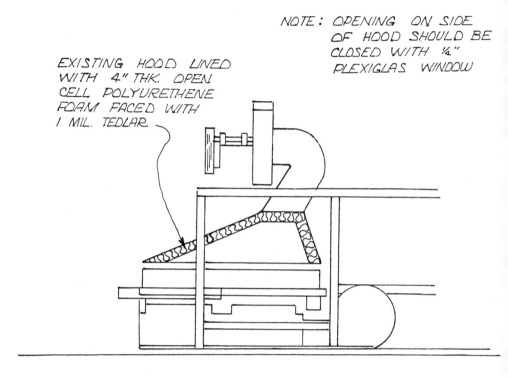

NOTE: OPENING ON SIDE OF HOOD SHOULD BE CLOSED WITH ¼" PLEXIGLAS WINDOW

EXISTING HOOD LINED WITH 4" THK. OPEN CELL POLYURETHENE FOAM FACED WITH 1 MIL. TEDLAR

Figure 14.1. Sound absorptive plenum for peanut sheller blowers.

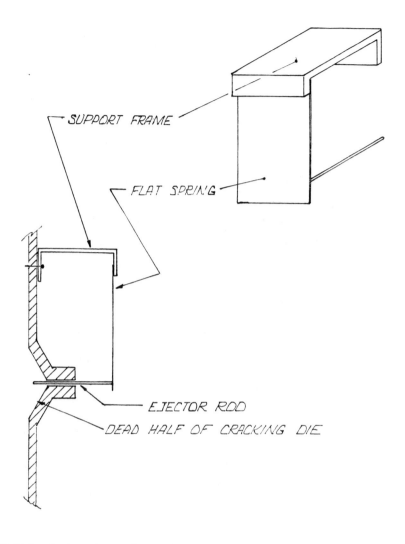

SUPPORT FRAME

FLAT SPRING

EJECTOR ROD

DEAD HALF OF CRACKING DIE

Figure 14.2. Spring ejector for crackers.

NOTE : FOR MACHINE ACCESS NOISE
       SHIELD IS LIFTED AND FOLDED ACROSS
       TOP OF NUT HOPPER.

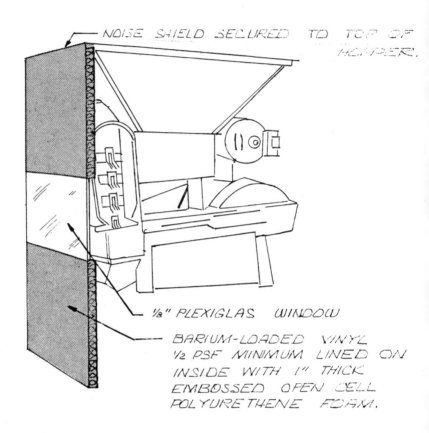

NOISE SHIELD SECURED TO TOP OF
                               HOPPER.

⅛" PLEXIGLAS WINDOW

BARIUM-LOADED VINYL
½ PSF MINIMUM LINED ON
INSIDE WITH 1" THICK
EMBOSSED OPEN CELL
POLYURETHENE FOAM.

Figure 14.3. Flexible noise shield for crackers.

15 - BAKERIES

In comparison with many other types of food plants, bakeries are relatively quiet. A typical bakery may have one or two areas where sound levels exceed the OSHA guidelines, impacting only a small percentage of the bakery employees. In a recent study published by OSHA, the estimated cost for noise abatement to bakeries was less than 1% of the total estimated to be required for all food plants. In noise surveys of four bakeries by our firm, each producing different types of products, the following acoustical environments were found:

   a. In a medium size bread bakery, sound levels in the dough making area were 93-96 dBA. The noise exposure levels in all other areas were within the OSHA limits, although employees were occasionally exposed to intermittent sound levels above 90 dBA.

   b. In a bakery producing cookies, sound levels due to normal operations exceeded 90 dBA at only one location, the pulverizer area. The unit was located in an isolated room, and employee exposure time to the 96 dBA sound levels was extremely low. It was also found that at only one other operation within the plant did the noise levels exceed 85 dBA. This was a cutting machine operation, and the noise was primarily due to old bearings which were replaced.

   c. In a bakery operation producing pie shells, excessive sound levels were identified in three locations; all three operations involved air noise. In two locations, the exhaust of pneumatic powered machines exceeded 95 dBA. The sound levels due to air used for ejection of the product out of pans also exceeded 90 dBA.

   d. In a plant producing frozen pizzas, excessive sound levels were measured in the pan washer area and the bake room. The highest sound levels in the plant were due to a bread grinder.

Were sound levels do exceed the OSHA limits, the bakery manager will find little consolation in the fact his noise problems may be less severe than those of other industries. Specific noise control solutions are required. The following approaches to noise control may be considered for various items of bakery equipment.

FLOUR HOPPERS

Excessive sound levels may be generated by the vibrations induced into the flour hoppers by air vibrators, as well as by the exhaust of the vibrator. These vibrations are, of course, necessary for proper material flow and cannot be eliminated. While maintaining the existing vibration level, the resulting sound levels can be reduced by acoustical lagging of the hopper.

The interior lagging treatment should consist of a 1" acoustical foam or glass fiber. The exterior of the acoustical foam should be wrapped with a 0.7-1.0 pound per square foot barium-loaded vinyl. The vibrator body should also be included in the wrapping, and only the inlet and outlet air hoses should penetrate the wrapping. All seams should be properly sealed to insure sanitation is maintained.

The use of new cushioned vibrators may be considered as direct substitutes for existing vibrators; however, these units often do not provide the same vibration characteristics. The exhaust of air vibrators always generates high noise levels and should be muffled. Electric vibrators may be used in place of pneumatic units to completely eliminate the exhaust problem.

An alternative method to the above would be to install an internal vibration system which would induce the vibration into the material directly, by-passing the hopper surface. Two schemes to accomplish this are outlined below. Either may be retrofitted to existing hoppers and will result in the elimination of hopper surface vibration as a noise source.

1. By placing a metal bar vertically in the hopper and vibrating it horizontally as a rigid body, the possibility of clogging would be decreased. Units of mild steel run from $2,000 to $12,000, depending upon the size of the hopper. Stainless steel units would naturally cost more. These vibrators have been successfully used to convey flour, and are available from:

    Vibra Screw, Inc.
    755 Union Boulevard
    Totowa, NJ 07511

2. By placing a screen of the same shape as the hopper interior inside the hopper and moving it to and fro, material would not be able to clog. The Thayer Bridge Breaker is such a device and is manufactured by:

    Thayer Scale
    Hyer Industries, Inc.
    Pembroke, MA 02359

    Stainless steel screens run from $800 to $1,600 per screen, depending upon the size and shape of the bin. It is not known whether these screens have been used to convey flour.

CONVEYORS

Two types of conveyors are most commonly used to convey bakery products from the dough making stage through ovens to final packaging: roller conveyors and woven wire conveyors.

The noise of metal pans on metal rolls may generate noticeable impact noise. This noise is generally less than 90 dBA for pans with product but often exceeds 90 dBA for empty pans. Three approaches to noise reduction may be considered:

1. Reduction of pan velocity (controlled by the incline angle of gravity feed conveyors).

2. Rubber coating of the rolls.

3. Substitution of the metal rollers with their plastic counterparts.

Woven wire conveyors are very quiet up to speeds of about 400 feet per minute, where a whine is observed due to the meshing of teeth with the belt. Such speeds are not encountered in bakeries. Noise may be encountered with the conveyor motor, and may be reduced as discussed in Chapter 23.

Additional noise reduction guidelines are presented in Chapter 16 for drive gears, chain conveyors, belt conveyors, and screw conveyors.

AIR EXHAUST

Noise control of air noise is presented in Chapter 24.

PUMPS

Pumps may be used, such as to pump dough from the mixer to a roll machine. Sound levels will depend upon the type of pump used. Gear driven, piston, and lobe type pumps may be quite noisy. Noise reduction of these units is difficult and may require enclosure or replacement of the pump. The quietest pump is the progressive cavity type. Noise associated with pump motors may be silenced as discussed in Chapter 23. Also see Chapter 19.

DOUGH CUTTING AND PRODUCT EJECTION

The thrust characteristics of air are commonly utilized for two functions in bakeries:

a. Ejecting product from pans.

b. Cleaning the edge of dough cut on Do-Makers.

A pressure of 40 psi is characteristic. The air is necessary to prevent a string of dough from trailing each cut from pan to pan.

Open air jets are typically used for these functions.

The thrust from an exit airstream is given by the equation:

$$T = \frac{W \cdot u}{g}$$

where: W = weight flow rate (lb/sec)
        u = airstream velocity (ft/sec)
        g = acceleration due to gravity = 32.2 ft/sec^2

The acoustic power (AP) in watts of an exit airstream is given by the relationship:

$$AP \propto \frac{Wu^7}{2gc^5}$$

where: c = velocity of sound (ft/sec)

With a seventh power dependency on air velocity associated with noise generation, but a direct relationship between thrust and velocity, an effective way to reduce high noise levels from nozzles is to obtain a nozzle which would exhibit somewhat lower exit velocity and reduce turbulence while maintaining high effective thrust. Several commercially available silencers have been developed to serve this function.

A recent study showed three commercially available silencers capable of producing thrusts of 32 ounces with sound levels from 5 to 10 dBA less than those associated with a free air jet of identical thrust.[1] These silencers are (in order of quietest first):

                Model: E-2-6 Ejector (1/4")
        Available from: C. W. Morris Company
                        36740 Commerce Street
                        Livonia, MI 48150

                Model: Type AE (1/4")
        Available from: Allied Witan Company
                        12500 Bellair Road
                        Cleveland, OH 44135

                Model: 309 (3/8")
        Available from: Sunnex
                        87 Crescent Road
                        Needham, MA 02194

Where there is not room for a muffler, such as on a Do-Maker, the use of a 1/4" AE type Collimator insert, available from Allied Witan (address above) may be considered.

It should also be recognized that localized shields which block the line-of-sight path between a high frequency noise source such as an air jet and an operator may reduce sound levels by approximately 5 dBA. Such shields may be constructed of transparent plastic materials.

PAN AND TRAY NOISE

The impacts of pans being loaded onto conveyors or being stacked may generate considerable noise, and presents a problem which is quite difficult to solve. Where loading is accomplished manually, operator technique is very important. It may also be feasible to develop non-metallic guides to reduce the velocity of pans before the impact occurs. Where pans impact metal stops, non-metallic materials may be installed (see Chapter 25).

PAN WASHER AREAS

In addition to the noise of pans, fan noise is generally present in washer areas. This noise may be reduced by means of fan inlet and outlet silencers.

DOUGH MAKERS

Several noise sources may be associates with dough makers:

1. In one bakery the drive unit generated sound levels of 95 dB at 250 Hz. This noise was reduced to 88 dB by enclosing the motor and drive unit.

2. Gears in the dough maker head induce vibrations into the housing, which in turn radiate noise. This noise may be contained by installing an enclosure or lagging treatment over the head. The lagging treatment would consist of a 1" thick glass fiber or foam in contact with the housing and completely covered by an isolated stainless steel skin. All joints should be scaled to prevent flour saturation. Existing grease fittings should be extended outside the panel to allow maintenance access.

3. Air may be used to prevent a string of dough from trailing each cut. Reduction of this noise was previously discussed in this chapter.

PANNER

Noise may be generated by the rattling of levers, drives and guides for the panner, divider, and associated equipment. This noise may potentially be reduced by replacing metal parts with impact resistant plastic parts in order to reduce rattle from metal-to-metal contact (see Chapter 25).

OVENS

Combustion noise in ovens is generally of low intensity and does not present noise exposure problems. The acoustic energy of the combustion process is broadband in nature, with peak energy in the 125-500 Hz range: the combustion roar frequencies. This frequency range relates to the flame propagation speed divided by flame thickness. Where sound levels are excessive, burner operation should be reviewed to identify if combustion oscillations are causing the problem. From measurements it is relatively simple to detect whether pulsations are present in the fuel and air supplies to a burner. These can be removed either by changing the fuel/air supply equipment, or fitting appropriate acoustic silencers.

In addition, potential abatement of combustion roar must be achieved by confining the noise to the firebox. It is not feasible to enclose the furnaces, since this would impede the air flow required for the combustion process; however, localized baffles may be considered.

GRINDERS

Sound levels of a bread grinder which exceeded 100 dBA were reduced by the installation of a quieter drive mechanism. Additional noise reduction could be achieved by the treatment of interior surfaces of panels of the housing with a damping treatment to reduce vibrations induced by the drive mechanisms. In some cases it may be possible to locate the grinder in an isolated room. Product may be fed by means of a conveyor.

PULVERIZERS

Sound levels of pulverizers may be as high as 100 dBA. In very few instances is constant attendance by operators required. Thus, location of pulverizers in isolated rooms or complete acoustical enclosures are the most direct approach to noise control. A case history involving one such enclosure by Ferro Corporation is summarized in Chapter 31. The use of localized enclosures may also be considered.

In addition to the direct radiation of sound, pulverizers will induce vibrations into the floor, which are transmitted throughout a building structure. These vibrations may annoy occupants of adjacent offices. This problem may be solved only by the installation of vibration isolators beneath the pulverizer.

MACHINERY NOISE

Mechanism noise may be identified in various items of bakery machinery, such as slicers, wrappers, etc. Generally, higher sound levels are associated with older or higher speed machinery. It is impossible to provide specific recommendations for such machinery, since specific noise sources will vary from machine to machine. Most frequently, this noise will be due to motors (Chapter 23), pneumatic controls (Chapter 24), metal-to-metal impacts (Chapter 25), or maintenance problems (Chapter 28).

MIXERS

Mixers are used to mix pastes and high viscosity fluids in the food industry. Designs vary depending upon types of fluids being mixed, their flow rate ratios, viscosity ratios and the non-Newtonian characteristics of the fluid. Applications vary from high to low shear rates, and running speeds may range from a few rpm to 1800 rpm.

Mixers generally do not present a major noise problem because:

    a. The direct noise of the process is contained in a bowl or drum.

b. Most mixers do not require constant operator attendance, reducing noise exposure time.

c. The paste or fluid serves to reduce impacts of the rotor-stator and dampen vibrations.

Where mixer noise levels are excessive, the following approaches may be considered:

1. The motors may be the dominant noise source and should be silenced (see Chapter 23).

2. The maintenance condition of the rotor, stator, and drive mechanism should be checked.

3. The structural rigidity of the mixer should be increased if the noise results from frame and panel vibration.

4. Where vibrations are due to panel resonances, vibration damping treatments should be applied.

5. If operator attendance is not required, the mixer should be isolated with a wall.

6. The mixing drum may be enclosed or lagged. The lagging treatment may consist of a 1"-2" layer of acoustical foam in contact with the drum, faced with an isolated exterior layer of 1.0 (minimum) stainless steel or barium-loaded vinyl. The foam, of course, must be completely sealed for sanitary purposes.

Vibrators are frequently for conveying materials and parts. The following approaches to noise control may be considered:

1. Install mufflers on pneumatic vibrators.

2. Substitute electric vibrators for noisier pneumatic units.

3. Use "cushioned stroke" vibrators.

4. Investigate alternate means of material handling.

5. Regulate vibrator use such that it is on only when required.

6. Operate vibrator at minimum required capacity.

7. Partially dampen system being vibrated.

8. Redesign system being vibrated to eliminate any large surface areas which would be prime sound radiators.

9. Enclose vibrator.

Noise control of vibratory bowl feeders is presented in Chapter 20.

CONVEYOR OPERATION

Guidelines for proper conveyor operation to insure minimum noise levels are presented in Table 16.1.

LIFT TRUCKS

An industry-wide (Industrial Truck Association) test procedure has been adopted which requires noise measurements to be made at the operator's ear plus 6, 12, and 18 feet from the side of the vehicle. These measurements are made at full speed, maximum load, and no load, plus during a "drive by". Many manufacturers specify sound levels in accordance with this standard.

Muffling of trucks may be accomplished by purchasing off-the-shelf mufflers and by shrouding the engine compartment. At present, fan noise is the major source of noise for LP gas vehicles, while high-speed DC motors are the major source of noise for electric vehicles. Power-steering pump noise also is a problem for the electric vehicles, but the noise of the electric vehicles is well below the OSHA requirements.

SCREW CONVEYORS

Screw conveyors can be significant noise sources. The mechanism of noise

generation is the structural vibration of the conveyor troughs. This pneno-
menon occurs only in those screw conveyors which are misaligned or which have
bent screws. Either of these conditions results in bearing failure and sub-
sequently screw support vibration. This vibration is transmitted to the
conveyor trough at the fastening points. The resulting noise is a high frequency oscillating screech. There are several ways that this problem can be
solved:

1. Replace bent screws, realign troughs, replace bad bearings, and then
   institute a program of regular maintenance to insure that these problems
   never again become significant.

2. Apply a surface damping treatment to all conveyor troughs.

3. Isolate the conveyor trough from the bearing excitation by inserting a
   low impedance material (such as rubber) between the bearing and conveyor
   trough as shown in Figure 16.1.

Reference

1. Wildsmith, C. G., "Better Operation of Conveyors and Elevators," *Food
   Manufacturer*, May, 1975, pp. 27-28.

TABLE 16.1

GUIDELINES FOR PROPER OPERATION OF CONVEYORS [1]

General

1. Are recommended lubrication procedures followed taking into account physical working conditions, intensity of operation and ambient temperature? And is the lubricant reaching the precise points where it is needed?
2. Is the equipment loaded in such a way as to avoid flooding with bulk materials and to reduce impact to a minimum, avoiding pulsations and surging?
3. Is the equipment always cleared before shutting it down? This avoids starting up under full load and resulting strain and wear.
4. Before starting up, is the equipment checked for obstructions or any undue build-up of materials – particularly those that tend to pack or harden – that could interfere with its operation?
5. Is the equipment run empty for a short period after starting up, to ensure fault-free operation and effective lubrication before being placed under full load?
6. Is the equipment run regularly – say once a week – during shut-down periods? This avoids risks of overloading through binding or seizing when the plant starts up again.
7. Are appropriate spares carried on site, to minimise downtime in the event of breakdown?
8. With outdoor installations in particular, is the supporting framework and casing, where fitted, kept painted against corrosion?

Drive gear

1. Are all the connected units correctly aligned? With V-belt drive, for instance, failure to ensure correct alignment of pulleys will result in rapid belt wear.
2. Is the drive gear inspected at least weekly for normal functioning without signs of overheating?
3. Is uniform tension applied to V-belts, with a slight 'bow' on the slack side of the belts when running under load?
4. Where oily conditions are encountered, are special V-belts fitted? Nor-
mally oil should not be allowed to come in contact with V-belts and belt dressing should never be employed on V-belts.
5. In the case of chain drives, is the slack side correctly tensioned? Over-tensioning will cause undue wearing of both chains and sprockets; too little tension may allow the chain to jump the sprocket teeth or the links to ride up the teeth, causing damage.

Chain conveyors

1. Is chain tension just sufficient to take up the slack and balanced where chains are used in parallel? Over-tensioning can create excessive wear.
2. Is the chain correctly aligned with the conveyor framework or casing and running centrally?
3. Is the chain's natural catenary interrupted at any point by rollers on the return strand? This can lead to high pin/bush wear.
4. Does the chain leave the driving and tail sprocket wheels cleanly without bunching between the wheels and chain support guides?
5. Are all rollers, where fitted, rotating freely and not sliding along their track supports?
6. Are chains used in sliding applications designed specifically for this type of duty?
7. Are sprockets of appropriate design, quality and not too small? The fewer the teeth the higher the 'chordal' action, producing increased pin/bush wear.
8. Are sprockets inspected regularly for signs of wear and to ensure correct meshing of the chain in the teeth? This affects gearing performances, jerky, irregular running causing rapid chain wear.
9. Are all attachment bolts securely tightened and damaged attachments replaced?
10. On overlapping slat or tray conveyors, does the slat profile allow sufficient clearance between the leading edge of the slat and the platform of the preceding slat? If not, or where other obstructions occur, severe stresses can be applied to the 'K' attachment of the chain – particularly where the return strands travel over rollers.

Belt conveyors

1. Are head and tail pulleys correctly aligned with each other and idlers also correctly aligned at right angles to the centre line of the belt, to ensure correct
tracking?
2. Has adjustment for correct tracking been carried out under loaded conditions, without undue reliance on 'knocking' or tilting of idlers, without unequal adjustment of take-up screws and without creating undue belt tension? All three of these methods of adjustment are incorrect and should be avoided.
3. Is applied tension just sufficient to prevent the belt slipping on the drive pulley under loaded conditions – and no more?
4. Are regular checks made to ensure that all idlers turn freely, any caked material being removed from idler and pulley surfaces?
5. Are belt ends joined by means of mechanical fasteners correctly cut (square with the centre line of the belt), checked regularly and fasteners replaced as necessary? Snapping of worn fasteners results in uneven strain that can tear the belt lengthwise.
6. Are any stationary parts of the equipment (other than belt scrapers) or material wedged beneath feed chutes allowed to come in scraping contact with the surface or edges of the belt?
7. Are any cuts and abrasions in belt covers or edges repaired promptly as they are noticed?

Screw conveyors

1. Are flights checked regularly for signs of deformation which could in turn lead to serious damage to troughs and/or drive units?
2. Are intermediate hanger bearings checked at least monthly for signs of wear? Abnormally worn bearings will lead to the spiral shaft becoming bent or distorted and in turn cause the flights to rub on the bottom of the trough.
3. If a gudgeon and flight section are removed for servicing or replacement, is the spiral shaft checked to see that it is free to be turned by hand before power is applied to the motor?

Chain and bucket elevators

1. Is the chain correctly tensioned just to take up the slack, with both chains equally tensioned in the case of double strand elevators? Incorrect tensioning where buckets are of the dredging type will cause the bucket position to become uncontrolled, increasing wear on sprockets, chain, boot casing and at worst jamming the elevator and breaking the chain.

TABLE 16.1 (Continued)

2. Are all bucket bolts securely tightened and overlapping buckets capable of moving freely?

3. Are skidder bars, where fitted, clearing the inside of the elevator casing, replaced if worn or broken, and are the skidder bar bolts securely tightened?

4. Is the chain inspected at least monthly for correct meshing in the sprocket wheel teeth and any signs of abnormal wear?

5. Are buckets fouling the bottom of the boot, due to chain wear or 'stretch'? If so, clearance must be maintained by removing a few links or a bucket pitch of chain.

6. Are damaged buckets replaced as necessary? This applies particularly to those of the overlapping type, where deformation will impair operation of the elevator and may cause jamming.

7. Are checks made to ensure that material is not bridging in the feed or discharge chutes and that, with dredging-type elevators in particular, material is not allowed to build up in the boot above the normal running capacity of the elevator? Flooding of the boot causes excessive wear and tension in the chain, overloading of the drive gear and possible jamming of the elevator.

8. In outdoor installations in particular, is the casing kept painted against corrosion? Remember that with freestanding elevators in particular the casing serves a structural role and corrosion, if unchecked, can have serious consequences.

Belt and bucket elevators

1. Is belt tension just sufficient to prevent the belt slipping on the head pulley when running under load conditions? Belt slip results in excessive belt wear, reduced elevator capacity and eventual choking of the boot if the normal rate of feed is maintained. Also, sagging of the return side of the belt can result in 'flapping', causing interference of the buckets with the elevator casing.

2. Are the take-up screws equally adjusted? If not, the belt will be thrown out of line.

3. Are the belt and buckets inspected at least weekly to ensure that the belt is correctly tensioned, belt joint fasteners secure, all bolts kept tightened, worn washers replaced, correct clearances maintained and any faulty buckets replaced?

4. Are steps taken to guard against flooding of the boot? This can cause damage particularly when the belt is started, straining the belt and often resulting in slackening of belt tension.

5. Are the buckets running clear of the boot?

6. Are checks made to ensure that material is not bridging in the feed or discharge chutes?

7. Is the elevator casing kept well protected with paint? (See point 8 under previous heading).

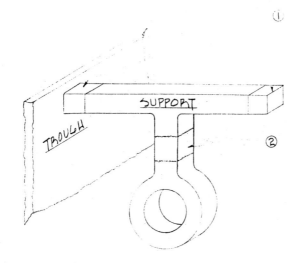

(1) Isolation achieved by separating support and trough with rubber

(2) Isolation achieved by making part of support neck out of rubber

Figure 16.1. Two methods of isolating the conveyor trough from bearing excitation.

17 - BEVERAGE BOTTLING AND CANNING

The noise associated with container impacts and equipment of beverage bottling and canning lines poses a major noise problem in the food industry. The problem is particularly severe in high speed and miltiple line operations, and is amplified by reverberation due to hard wall and ceiling surfaces. The following sections are designed to explain the mechanisms involved in noise generation of canning and bottling operations and to present solution approaches which may be applicable in many plants.

EMPTY CAN CONVEYORS

High background ambient sound levels are observed due to empty can conveying. Typically, a cable conveyor is used to transfer cans from the depalletizer to the filler room. Although the sound levels three feed from a conveyor are generally only 95-100 dBA, very little attenuation is observed with increased distance from the conveyor due to its length (an acoustic line source) and usual location near walls and ceiling (acoustic radiation into a quarter-space provides a 6 dB amplification). This source creates an ambient sound level throughout work areas which is typically near 90 dBA, with surges generating sound levels averaging 100 dBA. In operations involving multiple lines, the ambient sound levels may easily exceed 100 dBA. The noise may be identified as resulting from four sources:

a. Impact resulting from the "start-stop" motion of the cans on the line.

b. Sliding friction of can-against-can movement.

c. Sliding friction of the can against the guide rails.

d. Repetitive impulse from can impacts against other cans and the guide rails.

The following approaches to noise control may be considered:

1. Do not route conveyor tracks through work areas.

   This approach provides the ideal solution for new plants, however, may not be considered practical for existing facilities. Moving existing conveyor tracks is quite costly, and may require moving the depalletizer. In many plants, there simply may be no alternate routes.

2. Balance conveyor velocity to 1-2 cans per minute over the filler demand to minimize surges, or adjust the speeds of the conveyor sections to move the can surges into a warehouse or other area away from employees.

   This solution will provide adequate noise reduction for some single line operations where the sections of the can track are located a reasonable distance away from employees. Where only a slight noise reduction is required, this is the simplest and least expensive answer for noise control.

3. Install new modulated speed control system.

    A modulated cable conveyor system utilizes motion sensors, electronic controls, and DC motor drives in modulating line speed to provide a constant container supply, and keeps a space between cans, eliminating can-to-can contact. In addition to reduced noise levels, reduced container damage and maximum productivity may also be achieved. A complete modulated system costs approximately $20,000. Systems are commercially available from several firms, including:

    > Electro-Sonic Control
    > 517 North Elizabeth
    > Milton-Freewater, OR 96862

    A modulated conveyor system will not eliminate can-rail friction noise. This will not be a problem where only a few lines are involved; however, in plants with dozens of lines, ambient sound levels up to 96 dBA may occur due to friction noise alone.

4. Install plastic side rails.

    The theoretical noise reduction achievable by introducing a plastic (Teflon) to steel surface ($\mu = 0.07$) to replace a steel to steel surface ($\mu = 0.2$) is 4.5 dB. Noise reductions of 2 dBA have been reported due to the use of nylon side rails.

5. Enclose the can line.

    Due to the many mechanisms involved in can conveyor noise, enclosure of the can line is the only solution which will provide adequate noise reduction for many lines. Costs are only moderate unless many lines are involved. Special design considerations must be observed to reduce potential maintenance and cleaning problems. An overhead can line cover which achieved a 12-14 dBA noise reduction is shown in Figure 17.1. The enclosure was supplied by:

    > Body Guard, Inc.
    > 420 East fifth Avenue
    > Columbus, OH 43201

6. Install plastic or wooden support.

    A 2-3 dBA noise reduction has been reported by replacing the standard metal sheaves with a wooden cable track.

CAN DROPS

Where can drops, 180 degree turns, and "waterfalls" are present in can lines, excessive noise may be generated by can impacts. Noise reduction may be achieved by enclosing or redesigning this section of the conveyor. Also, rubber guides may be installed to reduce can velocity, such as shown in Figure 17.2.

FILLERS

Air is used to raise empty bottles for filling. When the bottle is filled, the cylinder mechanism is tripped at a stationary point and the air is released, generating noise.

The noise due to the filler cylinder discharges may be reduced by the installation of a box-shaped muffler at the discharge location, as shown in Figure 17.3. The front of the silencer should have a clear plastic door to provide cleanability, and the back should be lined with acoustical foam, as described above. Baffles provide air diffusion in the design. The top and bottom of the silencer are open.

The discharge noise may also be reduced by modifications to each air valve. Costs are estimated to be $30 per valve, but due to the number of valves involved (typically 60) and the downtime required for modification, a fixed location silencer is the preferred solution.

The relief valve of fillers may generate sound levels of up to 115 dBA and may be silenced using a conventional pneumatic silencer (see Chapter 24).

Significant noise is generated by the passage of air through the vent tube as a measure to minimize foaming. This is also known as a "scavenger". The only simple solution to this problem appears to be to review the necessity of blowing off the vent tubes, and if it is necessary, to minimize the valve opening. If this measure does not provide adequate noise reduction, the filler manufacturer should be contacted with regard to:

   a.   Changing vent tubes.

   b.   Achieving various settings in the production equipment to minimize foaming.

The impact noise of bottles feeding the filler may be reduced by means of an infeed worm screw.

The noise associated with bottle handling may be reduced by the use of non-metallic parts, as discussed in the next section of this chapter. The cost for replacement of starwheels and the A-frame will range from $1000 to $2000.

From the above discussion, it is seen that there are several mechanisms of noise generation involved in filler operation, and that noise reduction, while possible, is not easy. An alternate approach is to install a clear plastic sliding barrier in front of the filler. In one plant, a noise reduction from 106 dBA to 96 dBA was achieved by the installation of sliding Plexiglas doors. Only a slight production interference problem was reported; however, the doors were frequently required to be left open for considerable time periods when filling problems were encountered. A filler barrier designed and supplied by Body Guard, Inc., Columbus, Ohio is shown in Figure 17.4.

"SOFT" BOTTLE HANDLING EQUIPMENT

Non-metallic (plastic or phenolic) bottle feed mechanisms are commonly used for noise reduction. The theoretical noise reduction achieved by cushioning bottle impacts would be in proportion to the increase in material elasticity:

$$NR = 10 \log \frac{E_1}{E_2}$$

For structures of similar rigidity, the theoretical reduction of impact noise by substituting plastic (Nylon: $E = 4,000,000$ psi) for steel ($E = 29,000,000$ psi) would be 18 dB. In practice a lesser but significant noise reduction is achieved.

CROWNER

Air jets are used on crowners for two functions:

    a.   To assist flow in the hopper.

    b.   To move crowns along the feed chute.

The hopper air jets may be replaced by agitator springs for slow speed lines. At higher speeds, a low pressure air jet will be required to assist agitation.

The crown feed air jet may be silenced by a pneumatic muffler, such as an AE Type, available from:

               Allied Witan Company
               12500 Bellaire Road
               Cleveland, OH 44135

The silencer will provide a major noise reduction for free air jets (without crowns); however, high noise levels will be generated by turbulence as the air jet passes the crown edge. This noise cannot be eliminated by design, but may be isolated by means of a localized enclosure over the crown chute or by the installation of a clear plastic barrier in front of the crowner.

CAN SEAMERS

The noise of can seamers is generally the highest noise source in a beverage canning plant, typically measuring 100 dBA at the operator position.

The primary noise source is the impact of the code wheel upon the can lid. Significant noise is also attributed to the cover separator mechanism. In addition, a loud double impact may occur when the lid turrent is out of time.

Angelus Sanitary Can Machine Company designed an experimental lined stainless steel barrier to fit and enclose the cover separator above the cap feed plate and to partially enclose the cap feed turret below the cap feed plate. A baffle was added under the cap feed plate. The following noise reductions were reported:

Speed	Before	After
750 cpm	93.5 dBA	86 dBA
1250 cpm	99.5 dBA	91 dBA

A seamer enclosure designed by our firm is shown in Figure 17.5.

The replacement of the code wheel may be considered as a possible approach to noise reduction. New systems which paint the code information at high speed are quiet, however, cost approximately $15,000.

FILLER ROOM ACOUSTICS

A sound level build-up of 3-7 dBA is often found in filler rooms due to reverberation caused by hard wall surfaces. Recognizing sanitation requirements, it may be possible to install sound absorptive panels with a thin plastic facing for cleanability. (See Chapter 22.)

LABELERS

Noise studies of three labelers indicate that the following sound levels are typical:

Distance	Sound Level, dBA
1'	94
3'	92
5'	91
8'	90

There are numerous noise producing sources associated with labelers, and an extensive design study would be required to develop mechanism modifications for noise control. Such designs may also interfere with the machine operation. Enclosure of the machine is not considered feasible.

Observations of labeler operations indicate that the operator is not required to spend all of his time directly at the machine, and can observe his operation from a distance of a few feet at some times. The most effective method of noise exposure reduction would be to limit the exposure of the operator to a maximum of 6 hours per day within a 3' radius of the machine.

A line painted at this distance may assist in implementation of this operational control.

BOTTLE CONVEYORS AND COMBINERS

Noise is generated on conveyors and combining tables due to bottle-to-bottle and bottle-to-rail contacts. The following approaches may be considered for noise reduction:

1. Smoothen bottle flow.
2. Eliminate unnecessary bottle pile-ups.
3. Reduce bottle impact velocity.
4. Utilize plastic guides and rails.
5. Redesign transfer points.
6. Use cradlebelts.
7. Install acoustical line covers.

Typical line cover designs are shown in Figures 17.6, 17.7, and 17.8.

CRADLEBELTS

To eliminate impact noise due to bottle or can contact on conveyor lines, the use of cradlebelts may be considered. They are available from (see Chapter 31):

Product Diversification, Inc.
10832 Chandler Boulevard
North Hollywood, CA 91601

BOTTLE WASHERS

The sound levels generated as the bottles exit a washer may be significantly reduced by treating the guides with plastic. A remaining noise is the impact of the bottles as they drop approximately 1" onto arms. The impact sound level (slow meter response) is 100 dBA in front of the washer. This noise may be reduced by reducing the drop height of the bottle. The Law of Conservation of Energy indicates that the energy associated with the fall will be totally converted into other energy forms. Vibrational and acoustical power are the primary resultant energy forms. Thus:

$$I_a \propto wh$$

where: I_a = acoustical energy
w = bottle weight
h = drop height

Thus, the expected noise reduction would be directly proportional to the bottle drop height. To achieve a noise reduction of 10 dBA, the drop height should be reduced from 1" to 0.1".

This may be accomplished by:

   a. Moving the arms upward by adjustment or by spacers.

   b. Modifying the guides to allow the bottle to slide onto the arms.

It is reported that the top of the bottles may break if the guides are raised too high. It may be necessary to modify the machine slightly to prevent such breakage.

Noise is also generated by bottle contact at the washer discharge. The use of a barrier, as shown in Figure 17.9 may be considered for noise reduction.

INSPECTORS

Bottle inspectors are frequently exposed to excessive noise due to adjacent equipment. Reduction of their noise exposure may be achieved by:

   1. Isolation of adjacent equipment by a wall.
   2. Noise reduction of adjacent equipment.
   3. Enclosure of the inspectors.
   4. Rotation of the inspectors.

DROP PACKERS

The drop of cans or bottles into trays or cartons results in high impact noise levels. The use of a rubber pad on the drop plate provides an evident solution; however, only a 1-2 dBA noise reduction is generally achieved.

Some can packers do not involve dropping cans and are inherently quiet. One design involves forming the tray around the cans.

Bottle drop packers may be replaced with lowering head case packers with a pressure controlled infeed for inherently quieter operation.

The use of an acoustical barrier may be considered for noise control. In one plant, a reduction from 107 dBA to 96 dBA of bottle impact noise was achieved by the use of an open-front clear plastic slitted curtain supplied by:

>   Doug Biron Associates, Inc.
>   P.O. Box 413
>   Buford, GA 30518

The enclosure of an infeed can packer shown in Figure 17.10 provided a noise reduction from 104 dBA to 88-90 dBA. The noise sources were identified as:

   a. Can-can contacts and infeed surges.
   b. A pneumatic rocker arm.
   c. Packer mechanisms.

The enclosure was constructed of panels supplied by:

Body Guard, Inc.
420 East Fifth Avenue
Columbus, OH 43201

A drop packer enclosure design is shown in Figure 17.11.

SIDE VIEW

Figure 17.1. Overhead can line cover providing 12-14 dBA noise reduction
(courtesy of Body Goard, Inc.).

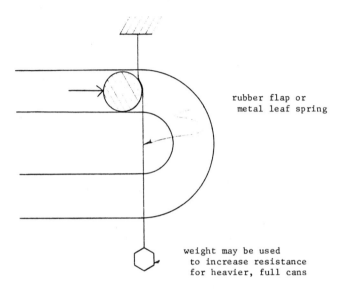

rubber flap or
metal leaf spring

weight may be used
to increase resistance
for heavier, full cans

Figure 17.2. Break to reduce can velocity on 180° conveyor turns or waterfalls.

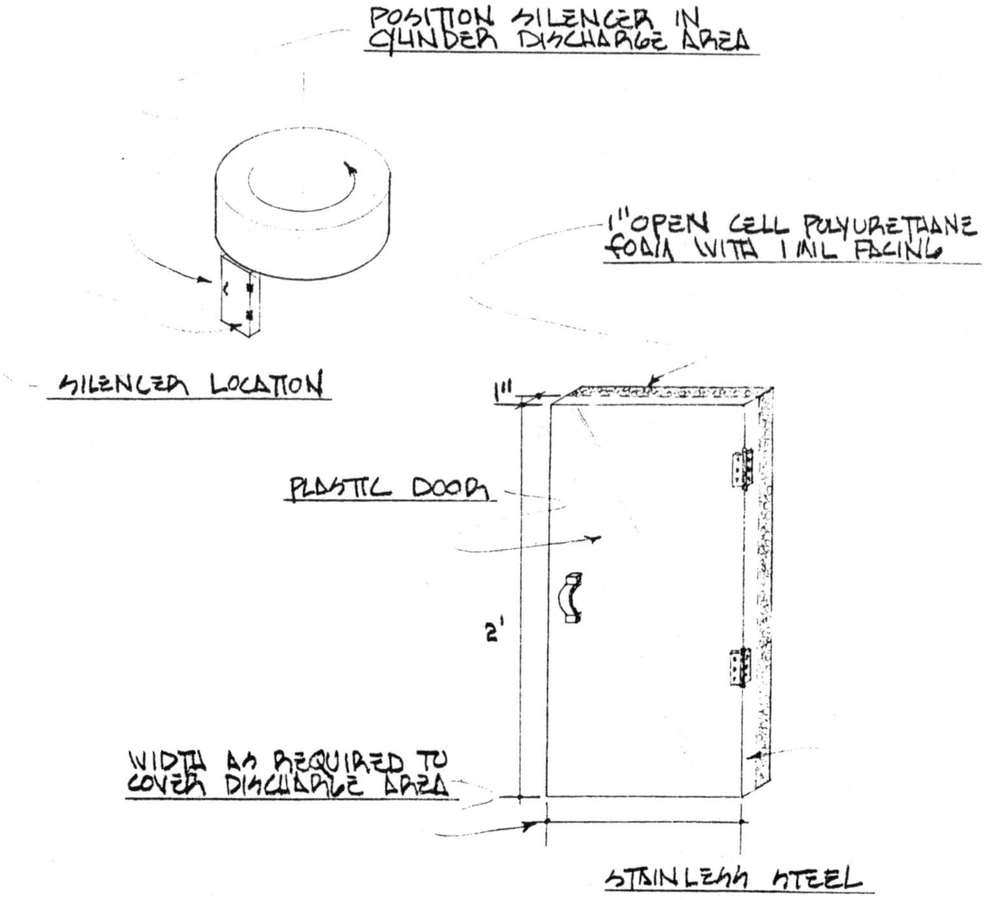

POSITION SILENCER IN CYLINDER DISCHARGE AREA

1" OPEN CELL POLYURETHANE FOAM WITH 1 MIL FACING

SILENCER LOCATION

PLASTIC DOOR

1"

2'

WIDTH AS REQUIRED TO COVER DISCHARGE AREA

STAINLESS STEEL

Figure 17.3. Silencer for filler cylinder discharge.

Figure 17.4.

Acoustical barrier for bottle
filler (courtesy of Body Guard,
Inc.).

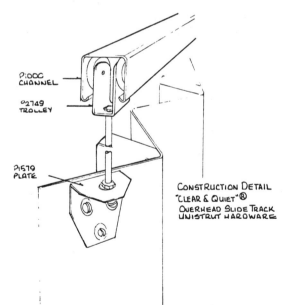

P1000
CHANNEL

O2749
TROLLEY

P1579
PLATE

Construction Detail
"Clear & Quiet"®
Overhead Slide Track
Unistrut Hardware

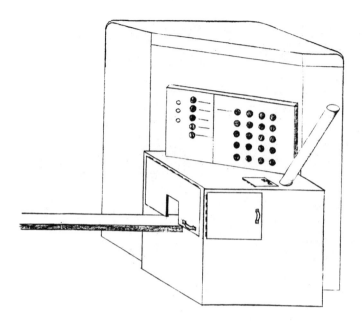

Figure 17.5. Enclosure for seamer openings.

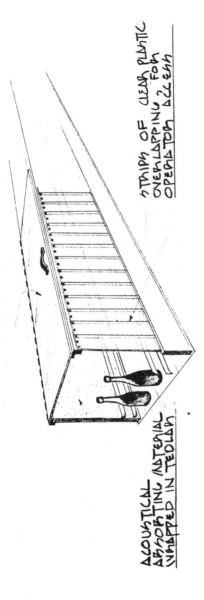

HINGED DOOR ⅛" (MIN.) PLEXIGLASS OR LEXAN

STRIPS OF CLEAR PLASTIC OVERLAPPING FOR OPERATOR ACCESS

ACOUSTICAL ABSORBING MATERIAL WRAPPED IN TEDLAR

Figure 17.6. Acoustical enclosure for bottle conveyor line.

-65-

Figure 17.7.

Acoustical line cover
(courtesy of Body Guard, Inc.).

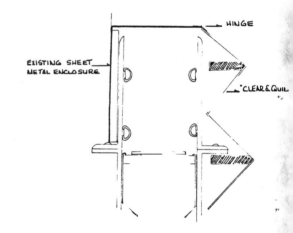

HINGE

EXISTING SHEET
METAL ENCLOSURE

CLEAR & QUIET

Figure 17.8. Traver covering, 8 dBA reduction (courtesy of Body Guard, Inc.).

CLEAR & QUIET® HINGED ACCESS

Figure 17.9. Bottle wash treatment (courtesy of Body Guard, Inc.).

Figure 17.10.   Infeed packer enclosure (courtesy of Body Guard, Inc.).

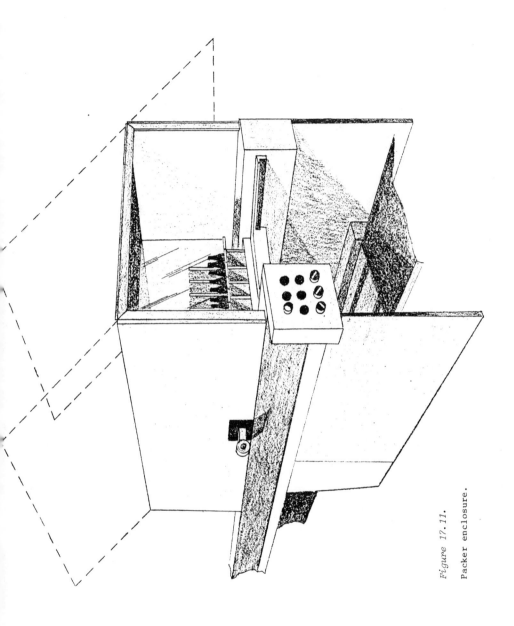

Figure 17.11.

Packer enclosure.

18 - TRUCK REFRIGERATION UNITS

The nighttime operation of truck refrigeration units can create a community
noise problem for dairies and other food plants. Two approaches to noise
abatement may be considered:

1. Acoustical barriers may be constructed to isolate the refrigeration
   unit noise from adjacent residences. Experimental noise reduction for
   acoustical barriers designed to isolate noise radiated from the front
   and rear of the units are shown in Figures 18.1 and 18.2. The barriers
   may be constructed to be portable on wheels, lift-on, or as a permanent
   structure similar to a car port.

2. Mechanical modifications may be made to the refrigeration units. In
   response to one community noise problem Thermo-King NWD units were modified by replacement of the fan and maintenance. The noise reduction
   achieved is shown in Figure 18.3.

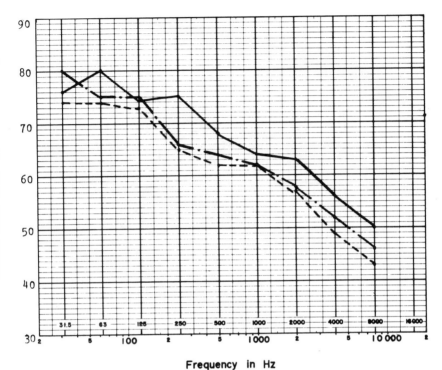

Figure 18.1. Octave band sound pressure levels for Thermo King NWD unit, measured at 50', for no barrier, and 4' x 8' barrier with and without 4" glass fiber facing.

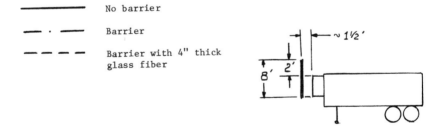

No barrier

Barrier

Barrier with 4" thick
glass fiber

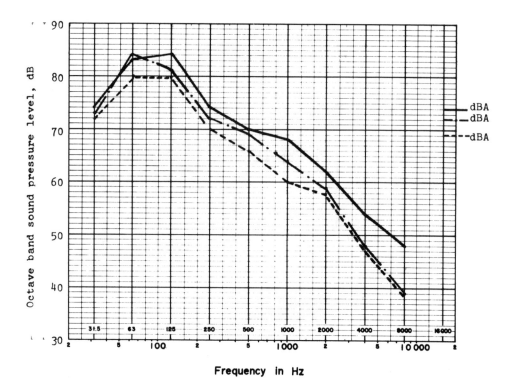

Figure 18.2. Octave band sound pressure levels, 20' to rear of NWD unit without barrier, and with 4' and 8' barrier.

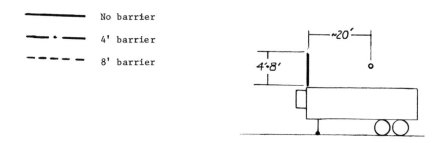

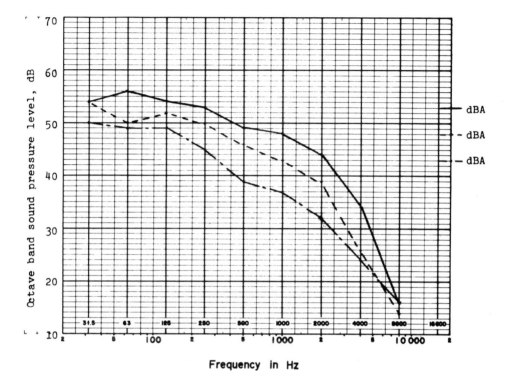

Figure 18.3. Octave band sound pressure levels of Thermo King NWD regriger-
ation units with and without modification, measured at
approximately 50'. The ambient sound levels are also
shown.

——————— NWD units

— — — — — Modified NWD units

——— · ——— Ambient

-73-

19 - FEED MILLS

The major noise sources in feed mills are flaking mills, hammer mills, and pellet mills.

FLAKING MILLS

The noise levels of soybean flaking roll machines have been measured to range from 93 dBA to 103 dBA. The primary mechanism of noise generation is structural vibration caused by the slight movement of the rolls as the product passes between them.

The following appraoches to noise control may be considered:

1. All openings on the front and top of the mill should be closed with a non-porous material weighing no less that 0.5 pounds per square foot. Suitable materials would include plywood, sheet metal, and barium-loaded vinyl. This feature will reduce the noise radiated outside the machine from the roller operation and the lower hoppers.

2. A sound absorbing panel of 2" thick open-cell polyurethane foam with a film of 0.1 mil Tedlar may be installed on the inner side of one of the sheet metal panels of the lower hopper, as indicated in Figure 19.1. The panel should be approximately ten square feet in area. The function of the panel is to reduce sound build-up within the machine structure.

3. The entire mill may be placed on vibration isolators to prevent vibrational energy transmission from the machine to the floor, lower hopper, and other structures.

4. The sheet metal panels of the lower hoppers may be treated with a constrained layer damping treatment, or may be constructed of commercially available vibration-damped panels.

5. The lower hopper may be installed with vibration isolation as indicated in Figure 19.2.

6. A lagging treatment may be applied to the entire face and back of the mill.

HAMMER MILLS

Sound levels of hammer mills may range up to 107 dBA. The dominant frequency of the noise is found to be at the hammer passage frequency:

$$f = \frac{n(\text{rpm})}{60}$$

where: f = frequency, Hertz
n = number of hammers
rpm = mill speed

The following approaches to noise control may be considered:

1. Use a lower speed motor (perhaps 1800 rather than 3600 rpm).

2. Enclose the mill with an absorbent-line box (see Figure 19.3).

3. Provide some means whereby the access doors on the upper inlet duct can be closed when access in not required.

4. Apply damping treatment to the exhaust duct.

PELLET MILLS

Sound levels ranging from 87 to 100 dBA are typical of pellet mills, and will be found to vary with operational conditions and the use of vibrators. The following approaches to noise control may be considered:

1. Vibrators should be used only when necessary.

2. Air vibrators should have exhaust mufflers.

3. Motor silencers may be installed (see Chapter 23).

4. The pellet mill gears may be treated.

5. An acoustical curtain may be installed on the pellet exit chute (see Chapter 26).

6. Gear guards and chutes should be dampened.

7. A rubber coupling may be installed in the drive shaft.

8. The die door may be treated with lagging or damping.

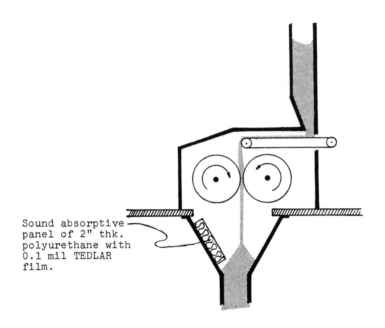

Sound absorptive
panel of 2" thk.
polyurethane with
0.1 mil TEDLAR
film.

19.1. Sound absorbant panels for interior of fl₊

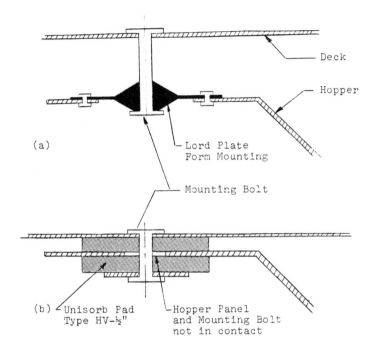

Figure 19.2. Two vibration isolation designs for lower hopper installation.

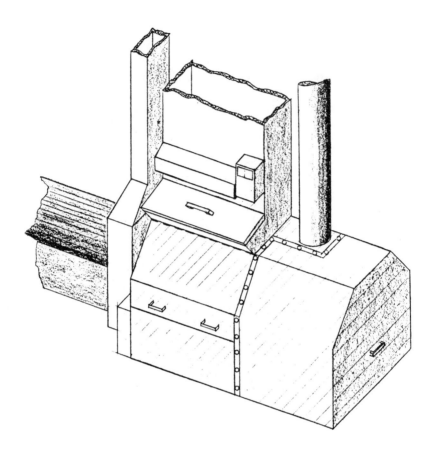

Figure 19.3. Enclosure for hammer mill.

20 - VIBRATORY BOWL FEEDERS

Vibratory bowl feeders are designed for a certain maximum design capacity, and are generally quiet within their design range. When this bowl speed is exceeded, excessive noise is generated due to system non-linearities. It is often observed that bowls are set to feed parts faster than required by the line flow. Maximum operating bowl speed should be determined for satisfactory operation and a limiter should be put on the potentiometer dial to prevent operation above this setting. The maximum operating point may be determined quiet easily by ear. If a limiter is not considered practical, the normal operational range should be marked on the potentiometer. In addition to noise level reduction, potentiometer control will result in both energy and machine wear savings.

Octave band sound pressure levels showing various modes of operation of vibratory bowls are shown in Figures 20.1, 20.2, and 20.3.

The sound levels of vibratory bowl feeders may easily be reduced to below 85 dBA by the installation of an acoustical enclosure. Two enclosure designs are shown in Figures 20.4 and 20.5. Noise control enclosures for vibratory bowl feeders are also available from the following manufacturers (cost generally ranges from $300-$800):

Ecology Controls, Inc.
223 Crescent Street
Waltham, MA 02154

Swanson-Erie Corporation
814 East 8th Street
Erie, PA 16512

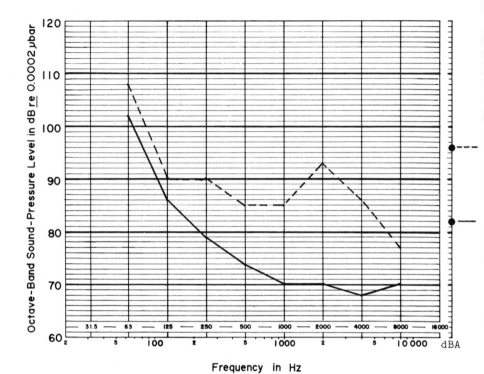

Figure 20.1. Vibratory bowl feeder without parts at normal potentiometer setting (————) and with potentiometer turned higher than design capacity (--------).

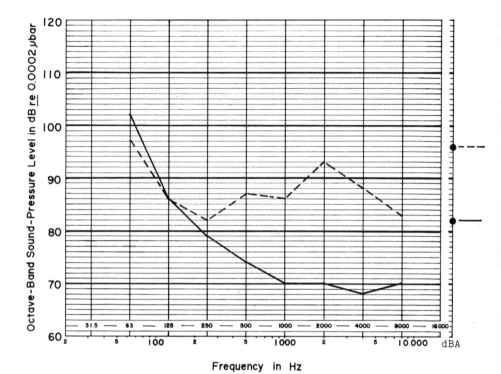

Figure 20.2. Vibratory bowl feeder empty (--------) and full (————) of parts.

-81-

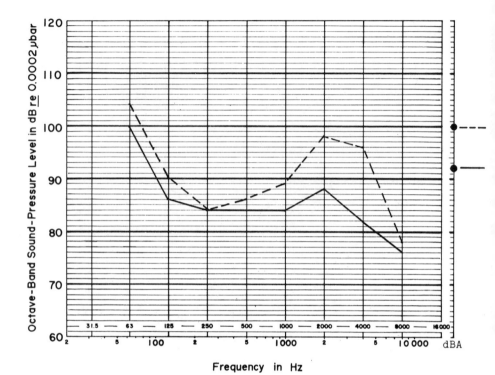

Figure 20.3. Vibratory bowl feeder during normal operation (‒‒‒‒‒‒‒‒) and
with cardboard cover (————). Measurement location 6" above
bowl.

1. Vibratory bowl feeder
2. Frame bolted to base for easy removal
3. 1.0 psf metal sides
4. Vibration damping material
5. 1" thick foam with 1 mil facing
6. Magnetic plastic gasket
7. Ferrous strip attached to cover
8. ¼" Plexiglas or clear acrylic

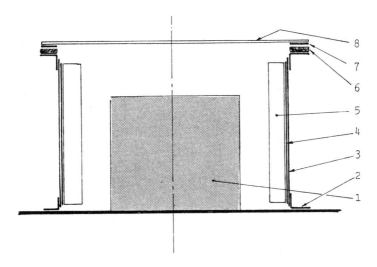

Note: Hinged and gasketed door may be
       used in place of cover.

Figure 20.4. Acoustical enclosure with cover for vibratory bowl feeders.

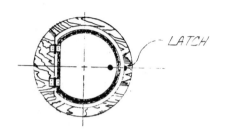

LATCH

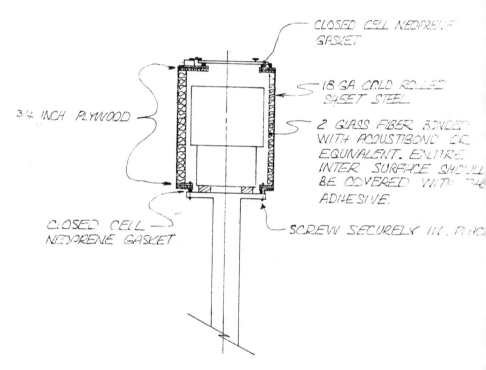

CLOSED CELL NEOPRENE GASKET

18 GA. COLD ROLLED SHEET STEEL

2 GLASS FIBER BONDED WITH ACOUSTIBOND OR EQUIVALENT. ENTIRE INTER SURFACE SHOULD BE COVERED WITH THE ADHESIVE.

3/4 INCH PLYWOOD

CLOSED CELL NEOPRENE GASKET

SCREW SECURELY IN PLACE

Figure 20.5. Acoustical enclosure for vibratory bowl feeder.

21 - VENTILATION FANS

The sound levels due to the roof ventilation fans often exceed 85 dBA, and occasionally exceed 90 dBA. Silencing of these fans is frequently required to achieve satisfactory ambient levels in a plant. Three approaches may be considered:

1. *Fan Silencers*. The installation of silencers is considered the most effective abatement method. The "straight-through" type should be used so as to minimize air flow impedence. Commercially available systems, such as from Industrial Acoustics Company, generally cost about $2000 per unit. Installation costs may be twice the unit cost.

2. *Replacement Blades*. There are commercially available replacement blades of a multitude of configurations for noise reduction purposes. Robertson Air Systems offers replacement blades that are compatible with most existing horsepowers, rpm's and sizes.

3. *Hanging Barrier*. Suspending a barrier directly below the fan, obstructing the line-of-sight to the workfloor may be considered. If a sound absorptive barrier of twice the fan diameter were hung 1½ times the diameter below the unit, a noise reduction of 1-4 dBA may be expected. A more restrictive barrier may provide noise reductions up to 10 dBA; however, such units are generally impractical, because a 10 to 15 percent impedance of the fan's air flow would result.

When purchasing new fan units, the engineer should always include noise level specifications.

22 - REVERBERATION

Whenever machinery is operated within enclosed spaces, sound levels will be increased to some extent due to reverberation. When this reverberant sound level increase becomes significant, it is appropriate to install sound absorptive materials on the ceiling above offending machinery. A simplified procedure can be used to estimate the increase in noise due to the reflected component, as shown in Figure 22.1.

Ceiling treatment is also required wherever acoustical barriers are employed to prevent sound from being reflected off the ceiling and over the barrier.

The most convenient method of employing sound absorption is the installation of acoustical baffles. The following is a list of manufacturers of commercially available baffle systems:

Owens Corning
Fiberglas Tower
Toledo, OH 43659

U.S. Gypsum
101 Wouth Wacker Drive
Chicago, IL 60606

Industrial Noise Control, Inc.
785 Industrial Drive
Elmhurst, IL 60126

Armstrong Cork Corporation
Lancaster, PA 17604

Eckel Industries, Inc.
155 Fawcett Street
Cambridge, MA 02138

Where fire sprinkler systems or air and ventilation requirements preclude the use of baffles, 1" thick conventional acoustical tile may be installed directly on the ceiling.

Where large surface areas are to be treated, a spray-on treatment may be most economical. These materials are available from:

National Cellulose Corporation
12315 Robin Boulevard
Houston, TX 77045

Sprayon Research Corporation
5701 Bayview Drive
Ft. Lauderdale, FL 33308

Attention should also be given to machine location within a room. These relationships are shown in Figure 22.2.

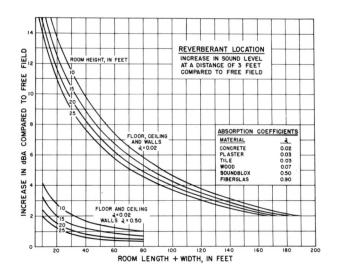

Figure 22.1. Simplified procedure for esitmating increase in noise due to reverberation.

Absorption Coefficient	Change Location	Increase In SPL
0.02	A to B	1.5
0.02	A to C	3.0
0.50	A to B	3.0
0.50	A to C	6.0

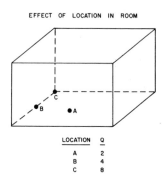

EFFECT OF LOCATION IN ROOM

LOCATION	Q
A	2
B	4
C	8

Figure 22.2. Effect on location in room.

23 - MOTORS

With approximately two million motors sold a year, motor noise is of great
concern. The two primary types of motor enclosures are the dripproof (DRPR)
and the totally enclosed, fan cooled (TEFC) motors. The DRPR cools itself
by pulling in outside air and circulating it around the electrical con-
ductors, while the TEFC motor uses a fan to accomplish cooling.

Table 23.1 presents a comparison of National Electrical Manufacturers Association (NEMA) and International Electrotechnical Commission (IES) motor
noise standards with those of a leading motor manufacturer. It is evident
that high speed (3500 rpm and above) and TEFC motors are noisier than their
slower speed and DRPR counterparts.

Solutions to motor noise problems include muting or muffling the inlet and
outlet air and isolating or dampening the vibrations. Motors with horsepower ratings of 75 or more often require abatement. Motors below this
rating do not normally generate noise levels that are above 90 dBA; however,
these smaller motors may contribute to other equipment noise, and the additive levels may exceed 90 dBA. The following firms manufacture motor silencers:

> The Spencer Turbine Company
> 600 Day Hill Road
> Windsor, CT   06095
>
> Pulsco Division
> American Air Filter Company
> 215 Central Avenue
> Louisville, KY   40201

Several motor manufacturers have developed new lines of quiet motors with
higher horsepower ratings. This has been accomplished through quiet fan
designs and minor internal modifications. To insure minimum motor noise
levels, specifications that sound pressure levels are not to exceed 85 dBA
at 3 feet and that sound power levels are not to exceed 95 dBA, should be
made to manufacturers for any new motors purchased. The following manufacturers have developed quiet motors:

> Westinghouse Electric Corp.   Reliance Electric Company
> Medium Motor & Gearing Div.   24701 Euclid Avenue
> Buffalo, NY   14240   Cleveland, OH   44117
>
> General Electric Company   The Louis Allis Company
> Schenectady, NY   12345   427 E. Stewart
>   Milwaukee, WI   53201

TABLE 23.1

A-WEIGHTED MOTOR SOUND POWER LEVELS
(dBA re 10^{-12} watt)

Horsepower	NEMA 3600 rpm DPPR	NEMA 3600 rpm TEFC	NEMA 1800 rpm DPPR	NEMA 1800 rpm TEFC	IEC 3600 rpm DPPR	IEC 3600 rpm TEFC	IEC 1800 rpm DPPR	IEC 1800 rpm TEFC	Lead 3600 rpm DPPR	Lead 3600 rpm TEFC	Lead 3600 rpm QUIET LINE	Lead 1800 rpm DPPR	Lead 1800 rpm TEFC	Lead 1800 rpm QUIET LINE
1			70	74		86		80				60	63	63
1.5	76	88	70	74		91		83	65	78	78	60	63	63
3	76	91	72	79	97	100	88	87	65	78	78	60	63	63
5	80	91	72	79	97	100	88	91	65	83	83	64	68	68
7.5	80	94	76	84	97	100	88	91	68	83	83	64	68	68
10	84	94	76	84	100	103	92	91	68	86	79	68	73	73
15	84	98	80	89	100	103	92	96	72	86	79	68	73	73
20	87	98	80	89	100	103	92	96	72	94	80	74	77	77
25	87	100	83	92	102	105	94	97	77	93	84	74	82	78
30	90	100	83	92	102	105	94	97	83	93	84	74	82	78
40	90	103	86	97	104	107	97	99	83	94	88	74	86	76
50	94	103	86	97	104	107	97	99	84	94	88	74	86	76
60	94	105	89	100	106	109	100	103	84	94	84	82	89	78
75	98	105	89	100	106	109	100	103	91	94	84	82	89	78
100	96	106	92	102	106	109	100	103	91	95	88	82	90	78
125	102	107	92	102	108	112	103	106	96	100	89	82	101	91
150	102	107	94	104	108	112	103	106	96	100	89	85	101	91
200	105		94	104	108	112	103	106	97			85		91
250	105				110	114	106	109	97					

TEFC: Totally enclosed fan
DPPR: Dripproof
QUIET LINE: Change of fan blade plus other motor properties

The generation of air noise results from the creation of fluctuating pressures due to turbulence and shearing stresses as high velocity gas interacts with the ambient air or solid surfaces. Radiating sources called "eddies" are formed with the high frequency noise being generated in the mixing shearing region and the lower frequency noise being generated downstream in the region of large scale turbulence.

Theoretically, as the pressure ratio between reservoir (line pressure) and ambient air is increased, the velocity of the air at the discharge nozzle increases. However, when a pressure ratio of approximately 1.9 is reached, the flow velocity through the nozzle becomes sonic, i.e., reaches the speed of sound, and further increases in reservoir pressure do not significantly increase the flow velocity. When this critical pressure ratio of 1.9 is reached, the nozzle is said to be "choked". It may be assumed, however, that at 80-100 psi, the air jets are generally in a condition of choked flow.

Based on the work of Lighthill, the overall sound power from a subsonic or sonic jet be calculated from:

$$P = \frac{K\rho A v^8}{c^5} \qquad (1)$$

$$
\begin{aligned}
\text{where:} \quad & A = \text{area of jet nozzle} \\
& \rho = \text{density of ambient air} \\
& v = \text{jet flow velocity} \\
& c = \text{speed of sound} \\
& K = \text{constant of proportionality}
\end{aligned}
$$

It is clear that the velocity of the gas stream has the greatest influence on jet noise. Cutting the velocity in half would lower the sound power by 24 decibels. However, halving the area would account for a decrease of only 3 decibels. For practical calculation of jet noise, Equation (1) may be modified to the following form:

$$P = \frac{eM^5 \rho V^3 A}{2} \qquad (2)$$

$$
\begin{aligned}
\text{where:} \quad & V = \text{average flow velocity through nozzle} \\
& M = \text{Mach number of flow (V/c)} \\
& \rho = \text{density of ambient air} \\
& A = \text{nozzle area} \\
& e = \text{constant of proportionality of the order of } 10^{-4}
\end{aligned}
$$

For choked flow (Mach 1) conditions, the factor eM^5 is approximately 3 x 10^{-5}.

The sound power from equation (2) may be converted to decibels utilizing the relationship:

$$L_w = 10 \log \frac{P}{10^{-12}}$$
(3)

The frequency of peak noise level can be calculated to the first order from the following equation:

$$f_o = \frac{SV}{D}$$
(4)

where: S = Strouhal number (a constant) = 0.2 approx.
for a wide range of conditions
V = nozzle exit velocity
D = nozzle diameter

Air discharges are observed throughout the plant and are considered a primary source of excessive sound levels in most areas. Most air discharges result from pneumatic exhaust of control systems, and these air exhausts are either continuous or cyclic in nature.

Sound levels generated by air exhaust, where concentrated air flow is not required, may be silenced by the installation of pneumatic mufflers which diffuse the air stream. The specification of an optimum pneumatic silencer for any application should be based upon:

· Sound level reduction
· Low pressure drop
· Durability
· Non-clogging features for the air contaminants present
· Economy

Mufflers should be periodically inspected for wear and effectiveness. If necessary an in-line filter system may be installed to prevent muffler clogging.

The following are manufacturers of pneumatic exhaust mufflers:

Permafilter
Div. Bonded Products, Inc.
P.O. Box 263
Wrentham, MA 02093

The Aeroacoustic Corp.
P.O. Box 65
Amityville, NY 11701

Air Noise

Allied Witan Company
12500 Bellaire Road
Cleveland, OH 44135

The following information should be provided to pneumatic muffler suppliers to assist in proper muffler selection:

  a. Pipe thread size (N.P.T.).

  b. Estimated or measured air flow (cfm).

  c. Presence of air line contaminants (oil, excess moisture, etc.).

It should be pointed out that most air discharges have threaded outlets which would easily accept silencers. To insure that the silencers are not removed, they may be secured by means of welding or a set screw. Signs reminding employees that noise control devices are for their benefit may also help.

Air leaks from pneumtaic systems are often found to be major contributors to the overall noise in many plants. It is important that a program be implemented including an inspection and maintenance procedure, performed on a regular basis, to identify air leaks such as due to the following:

  • Broken air hoses and cracked pipes
  • Worn fittings and couplings
  • Faulty valves
  • Air-operated devices left on when not in use

It should also be noted that such a maintenance program would be a significant cost and energy conservation measure, as well as an important part of a noise abatement program. The air consumption and cost associated with air leakage from a sharp-edged orifice continuously at 100 psig with air at 7 cents per 1000 cubic feet is as follows:

Diameter of Opening-Inches	Air Flow, cfm	Annual Cost of Waste (8 hour shifts)	Annual Cost (continuous)
1/32	1.62	$14.19	$59.60
1/16	6.49	$56.85	$238.78
1/8	26.0	$227.76	$956.59
1/4	104.0	$911.04	$3826.37

25 - METAL-TO-METAL IMPACT NOISE

Where structures or parts are impacted with metal-to-metal surface contact, a large portion of the impact energy is converted to vibrational energy, and in turn sound. The acoustical energy generated by part impacts is directly proportional to the kinetic energy at the impact point.

$$I_a \propto K.E. = \tfrac{1}{2}mv^2$$

where: I_a = acoustical energy
$K.E.$ = kinetic energy
m = part mass
v = part velocities

One method of reducing impact noise is to modify the system to reduce the impact velocity. Another method is to interrupt the metal-to-metal contact with a cushioning material, which serves to reduce the momentum transfer of the impacting structure.

Four types of impact surfaces may be considered, and it is found that the strongest of the materials will offer the least isolation. Selection of an optimum material must be made on an experimental basis. Three classes of materials to be considered are as follows:

1. The highest degree of impact isolation is offered by a rubber surface. It is found that rubber will wear very well in many situations, but is unacceptable in others. Rubber materials with good wear properties are:

Material	*Manufacturer*
Trellex 60	Trelleborg Rubber Company, Inc. 30700 Solon Industrial Parkway Solon, OH 44139
Armaplate and Jade Green Armabond	Goodyear Tire & Rubber Company Industrial Products Division Akron, OH 44136
Metalbak	Linatex Corporation of America Stafford Springs, CT 06076

2. Of great interest for highly stressed mechanical components are the plastics and their characteristics listed in Table 25.1. We have contacted several major manufacturers of plastics and find the following products available for impact isolation:

Metal-to-Metal Impact Noise

Product	*Manufacturer*
Lexan	General Electric 1 Plastics Avenue Pittsfield, MA 01201
Zytel ST 801	E.I. DuPont de Nemours Co. Plastics Department 170 Mount Airy Road Basking Ridge, NJ 07920

3. An impact energy absorbing foam, C3002-7, is available from:

Doug Biron Associates
P.O. Box 413
Buford, GA 30518

This material may also be faced with an exterior layer of damped sheet metal to serve as a protective facing.

TABLE 25.1

TYPICAL PLASTIC MATERIALS FOR IMPACT CONTROL

A. *Nylon.* For general purpose gears, mechanical components; has good vibration damping, machines well, resists oils and solvents.

B. *Acetals.* For accurate parts, maximum fatigue life, good machineability and resistance to oils, solvents and alkalis.

C. *TFE-Fiber Filled Acetals.* For heavy duty applications, excellent wear life, creep resistant, self-lubricating, low friction, good machineability and resistance to oils and solvents.

D. *Polycarbonates.* For intermittent very high impacts (not for repeated cyclic stress), creep resistance, dimensional stability, machines well, resists acids.

E. *Fabric-Filled Phenolics.* For low cost stamped gears or parts, creep resistance, good mechanical strength, resistance to oils and solvents.

F. *Glass-Filled Phenolics.* For highest mechanical strength and temperature resistance.

G. *Glass Fibric Epoxy.* For highest electrical and mechanical properties.

H. *Polyester.* High impact.

Reference

Mazoh, M., "Modern Materials for Noise Control," Design Engineering Conference, May 10, 1972.

26 - ACOUSTICAL CURTAINS

Flexible curtains of mass-loaded vinyl have excellent sound transmission loss properties and may be used to block airborne sound. Typical applications include:

- Complete or partial machine enclosures.

- Moveable walls to isolate noisy machine areas from other quieter areas.

- Localized enclosures for machine parts.

Acoustical curtains are constructed of 0.5 to 1.0 psf lead-filled vinyl and may be installed with grommets or on sliding tracks. It is important that the curtains extend to the floor and be sealed at the edges with Velcro fasteners for maximum sound attenuation. Installation details of enclosure construction are given in Figure 26.1. The following manufacturers provide acoustical curtains:

> Armstrong Cork Company
> Lancaster, PA 17604
>
> Doug Biron Associates
> P.O. Box 413
> Buford, GA 30518
>
> Ferro Corporation
> 34 Smith Street
> Norwalk, CT 06852
>
> Industrial Noise Control, Inc.
> 785 Industrial Drive
> Elmhurst, IL 60126
>
> Singer Partitions, Inc.
> 444 North Lake Shore Drive
> Chicago, IL 60611

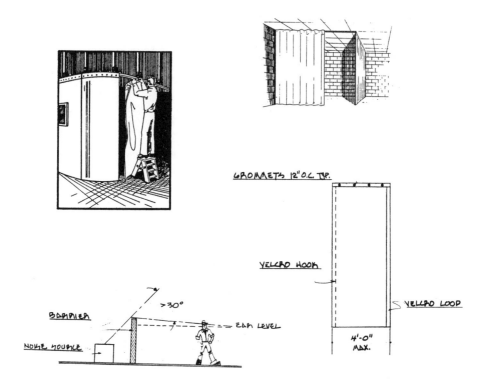

GROMMETS 12" O.C. TYP.

YELCRO HOOK

YELCRO LOOP

BARRIER

NOISE SOURCE

>30°

EAR LEVEL

4'-0" MAX.

Figure 26.1. Installation guidelines for acoustical curtains.

27 - EMPLOYEE ENCLOSURES AND BARRIERS

Enclosures, partial enclosures, and barriers are often considered as measures to reduce employee noise exposure from adjacent operations. These measures may be quite effective in some instances, however, may be ineffective or actually increase an operator's noise exposure in other circumstances. Two general guidelines should govern enclosure and barrier application:

1. An employee enclosure is effective only when an employee's job tasks allow him to spend a significant portion of his workday in an enclosure.

2. An acoustical barrier is effective only when the receiver is in the direct field rather than reverberant field of a noise source.

If a four-sided enclosure is applicable, windows, ventilation, and communication equipment should be installed. This type of booth can be used where the operator does not rely on audible detection techniques for operation. This kind of enclosure is ideal for control rooms and for rest area locations where there are often employees present who are not directly involved in the area operations and are present for environmental benefits.

A three-sided, or lean-to type enclosure is applicable where only a small noise reduction is required. Such an enclosure provides safety as well as production advantages. In operations which require that the operator be able to hear the machine, this type of enclosure may reduce sound levels to an acceptable level while permitting effective audible monitoring of the operation. The enclosure may have several openings or exits for accessibility and minimization of potential entrapment hazards, and can be equipped with windows as required. The following is a list of manufacturers of commercially available enclosures:

>Keene Corporation
>2319 Grissom Drive
>St. Louis, MO 63141
>
>Industrial Acoustics Company, Inc.
>380 Southern Boulevard
>Bronx, NY 10454
>
>Doug Biron Associates
>P.O. Box 413
>Buford, GA 30518
>
>Eckel Industries, Inc.
>155 Fawcett Street
>Cambridge, MA 02138
>
>Ross Engineering
>P.O. Box 751
>New Brunswick, NJ 08903

Employee Enclosures and Barriers

Sound Fighter Systems
1200 Mid-South Towers
Shreveport, LA   71101

An acoustical barrier, where applicable in an industrial environment, would
seldom provide more than a 3-7 dBA noise reduction. To achieve this reduction, the barrier must completely block the line-of-sight path between a
noise source and the observer. The barrier material must be without holes
or openings, and should be of 0.5 pound per square foot minimum weight. The
distance from a noise source where the reverberant sound field is equal to
the direct sound field may be computed from:

$$r = 0.14\sqrt{\bar{\alpha}S}$$

where: r = distance from source, ft
        $\bar{\alpha}$ = average sound absorption coefficient
        S = total surface area of the interior space, ft^2

For an observer located at this distance from a noise source, a barrier may
provide a 3 dBA maximum noise reduction. A barrier would be ineffective
where an observer is located a greater distance from the noise source.

28 - ACOUSTICAL MAINTENANCE

Machine inefficiencies, wear and malfunction can result in significant increases in noise levels. The goal of normal plant maintenance is to keep machinery in proper operating condition for efficient production. By simply expanding the existing maintenance program through noise awareness, it is possible to minimize the noise environment within the plant. If noise criteria are not assigned a level of priority, a machine with noise-producing maintenance problems, but which is still operating at 100% efficiency, would not justify maintenance attention. Because of this fact, acoustical maintenance may be met by the same opposition as its counterpart, preventive maintenance; cost savings are often hidden rather than direct.

Acoustical maintenance (AM) must be viewed as a separate discipline and an added responsibility to normal maintenance procedures. Noise control through maintenance would require an educational program for maintenance personnel to create "noise awareness" and outline engineering basics similar to those employed for preventive maintenance. This will provide maintenance personnel with the technical background necessary to cope with the unique engineering features associated with noise reduction. An effective program would involve creating a "noise awareness" among all personnel, and designating individuals whose primary duties are those related to plant-wide noise reduction. These duties may also include general energy conservation. A training program for maintenance personnel may follow the following outline:

    I.   Introduction to Noise Control
         A.   Decibels
         B.   OSHA
         C.   Sound Level Meters
         D.   Materials

   II.   Machine Design and Noise Control
         A.   Air Sources
         B.   Mechanical Sources

  III.   Application
         A.   Inspection Procedures
         B.   Guidelines for Installation

The importance of noise control related to maintenance cannot be over-emphasized. In a survey of 195 machines, it was found that over 60% were operating with a sound level at least 3 dBA louder than should be expected for a well-maintained machine. In one case, two identical machines were observed to have sound levels of 84 dBA and 107 dBA due to maintenance problems in one machine.

To organize an acoustical maintenance program, a maintenance engineer should be trained in noise control. Following this training, his duties would be as follows:

1. The engineer would perform periodic inspections to identify noise prob-
    lems related to maintenance.

2. He would report noise problems to the the maintenance department to
    schedule for repair.

3. The engineer would specify and order any acoustical materials not in
    stock.

4. He would consult with maintenance personnel if technical questions
    should arise regarding implementation.

5. He should inspect the repairs upon completion.

6. The engineer should maintain records regarding noise control inspec-
    tions and repairs.

The first task of the engineer is to document sound levels of each item of
machinery when in good maintenance condition. This data serves as a baseline to identify when excessive noise is present during future surveys.
Following the development of baseline data, noise survey inspections should
be performed periodically (typically at one month intervals), and machine
problems which cause measured sound levels to be 2-3 dBA higher than the
baseline data should be identified and reported for repair. In addition,
careful visual inspection should be made of each machine during each inspection. Particular attention should be given to:

> • Alignments and adjustments
> • Vibration and impact treatments
> • Air systems
> • Lubrication
> • Machine dynamics
> • Acoustical installations

29 - FIRE CODES

Manufacturers of acoustical materials and the Insurance Services have adopted a system using the flame spread classification (FSC) to rate combustible and non-combustible materials. The classification system is as follows:

Class	Flame Spread (FSC)	Characteristic
1	25 or less	Non-combustible
2	26 to 75	Combustible
3	76 & above	Combustible

Underwriters Laboratories, Factory Mutual Research, and Southwest Research Laboratories test and rate materials on a client basis under ASTM E-84. The U.L. 723 tunnel test, approved as ANSI A2.5-1970, is synonymous with the E-84 test method. Other standards for fire resistance rating of wall, floor, and ceiling materials are presented in U.L. 263 and ASTM E-119. Also, fire ratings for doors and wall systems are commonly presented on a time basis.

The engineer should insure that acoustical installations meet fire code standards, or plant safety may be jeopardized, and insurance rates may be raised.

Building construction is under the jurisdiction of the National Fire Protection Association. Local codes may add extra supplemental regulations without altering the basic specification. The local inspectors play a major role in the acceptance of materials installed. Insurance rates are readjusted according to compliance with national fire codes, nature of occupancy, and the fire protection class, rated 2 to 10 (best to worst respectively). Rates may be affected up to 100%.

In the case of enclosures with foam lining, the running footage of the building is calculated. The enclosure area is considered as a partitioned wall, and its running footage is also calculated. If the footage of the enclosure is less than 25% of the overall, there is no rate adjustment. If the footage is 25% to 50%, a 10% increase can be expected. If the percentage is 50% or more, the increase would be 20%; if it is 100% or over, there is a 40% increase. Where footage equals or exceeds 200%, a 100% rate increase is applied.

Generally, use of foam in a building, especially vertical applications, would change a non-combustible metal building rating (NC-2) to a combustible frame building rating (wood). A similar condition would occur where combustible spray-on acoustical treatment is utilized. The rating would change from a metal building to a frame building. For example, in a protection class of 3, the base rate of $.125 for a metal building would increase to $1.90, the rate for a frame building. Ceiling-hung combustible baffles or ceiling tile may increase rates 50% to 100%, depending on the amount of concealed area. Trowel-on dampening, a material non yet investigated, would fall under a rate schedule for hazardous conditions. Depending on the severity of the conditions, the rates could increase from 50 to 400%.

Generally, changes from a metal, non-combustible building rating to a masonry rating can increase rates on an average of 100 to 150%, and further, changes to a frame building rating (wood, combustible) can increase the rates an average of 150 to 200%.

A letter requesting a list of approved materials and/or fire ratings was sent to over 200 manufacturers of acoustical materials. Sixty-eight percent of those responding did not present fire ratings of any type in the consumer-available literature. Table 29.1 presents a list of all acoustical manufacturers in the United States known to provide fire ratings for their materials and systems.

In some cases acoustical materials have not been tested due to the fact that their classification is governed by the system for which they are used. For this reason, it is difficult to rate some acoustical materials, such as trowel-on dampening, since they are a part of an integral system, and the industrial machines for which they are used have not been considered for classification at all. There are acoustical systems, such as composite panels of perforated sheet metal, foam or glass bonded to sheet metal, which are commonly used, and still have not been listed.

TABLE 29.1

MANUFACTURERS OF FIRE-RATED
NOISE REDUCTION PRODUCTS

MANUFACTURER	PLYWOOD	GLASS FIBER	ACOUSTICAL CEILING SYSTEMS	DAMPING	SPRAY ABSORPTION	DOORS	SILENCERS/MUFFLERS	GYPSUM BOARD/PARTICLE BOARD	PLASTICS	LEAD	FOAM	CONCRETE BLOCK/CERAMICS	LEAD-LOADED VINYL/LOADED VINYL	METALS	ACOUSTICAL WALL SYSTEMS	ACOUSTICAL PANELS
Aeroacoustic Corporation 1465 Strong Avenue Copiague, NY 11726 (516) 226-4433		•					•									
Air-O-Plastik Corporation Asia Place Carlstadt, NJ 07072 (201) 935-0500		•										•				
Alpro Acoustics Division Structural Systems Corporation P.O. Box 30460 New Orleans, LA 70190 (504) 522-8656			•												•	•
American Smelting & Refining Co. 150 St. Charles Street Newark, NJ 07101 (201) 589-0500										•						
Arrow Sintered Products Company 7650 Industrial Drive Forest Park, IL 60130 (312) 921-7054							•									
Brunswick Corporation 1 Brunswick Plaza Skokie, IL 60076 (312) 982-6000															•	

TABLE 29.1 (Continued)

MANUFACTURER	PLYWOOD	GLASS FIBER	ACOUSTICAL CEILING SYSTEMS	DAMPING	SPRAY ABSORPTION	DOORS	SILENCERS/MUFFLERS	GYPSUM BOARD/PARTICLE BOARD	PLASTICS	LEAD	FOAM	CONCRETE BLOCK/CERAMICS	LEAD-LOADED VINYL/LOADED VINYL	METALS	ACOUSTICAL WALL SYSTEMS	ACOUSTICAL PANELS
Certain-Teed Products Corporation CSG Group Valley Forge, PA 19481 (215) 687-5500		•														
Conwed Corporation 2200 Highcrest Road Saint Paul, MN 55113 (612) 645-6699			•													
Doug Biron Associates, Inc. P.O. Box 413 Buford, GA 30518 (404) 945-2929			•	•					•	•		•		•		
Ferro Corporation 34 Smith Street Norwalk, CT 06852 (203) 853-2123												•	•			
Globe Industries, Inc. 2638 E. 126th Street Chicago, IL 60633 (312) 646-1300														•		
Gypsum Association 201 North Wells Street Chicago, IL 60606 (312) 491-1744	•							•								

TABLE 29.1 (Continued)

MANUFACTURER	PLYWOOD	GLASS FIBER	ACOUSTICAL CEILING SYSTEMS	DAMPING	SPRAY ABSORPTION	DOORS	SILENCERS/MUFFLERS	GYPSUM BOARD/PARTICLE BOARD	PLASTICS	LEAD	FOAM	CONCRETE BLOCK/CERAMICS	LEAD-LOADED VINYL/LOADED VINYL	METALS	ACOUSTICAL WALL SYSTEMS	ACOUSTICAL PANELS
The Harrington & King Perforating Co. 5655 Fillmore Street Chicago, IL 60644 (312) 626-1800														•		
Holcomb & Hoke Manufacturing Co., Inc. P.O. Box A-33900 Indianapolis, IN 46203 (317) 784-2444							•									
Koppers Company, Inc. Pittsburgh, PA 15219 (412) 319-3300	•															
Metal Building Interior Products Co. Lakeview Center 1176 E. 38th Street Cleveland, OH 44114 (216) 431-6040		•	•													•
National Cellulose Corporation 12315 Robin Boulevard Houston, TX 77045 (713) 433-6761					•											
Nichols Dynamics, Inc. 740 Main Street Waltham, MA 02154 (617) 891-7707		•													•	

TABLE 29.1 (Continued)

MANUFACTURER	PLYWOOD	GLASS FIBER	ACOUSTICAL CEILING SYSTEMS	DAMPING	SPRAY ABSORPTION	DOORS	SILENCERS/MUFFLERS	GYPSUM BOARD/PARTICLE BOARD	PLASTICS	LEAD	FOAM	CONCRETE BLOCK/CERAMICS	LEAD-LOADED VINYL/LOADED VINYL	METALS	ACOUSTICAL WALL SYSTEMS	ACOUSTICAL PANELS
Pittsburgh Corning Corporation Geocoustic Systems 800 Presque Isle Drive Pittsburgh, PA 15239 (412) 261-2900			●													
The Proudfoot Company, Inc. P.O. Box 9 Creenwich, CT 06830 (203) 869-9031												●			●	
Scott Paper Company Foam Division 1500 E. 2nd Street Chester, PA 19013 (215) 876-2551											●					
Singer Partitions, Inc. 444 N. Lake Shore Drive Chicago, IL 60611 (312) 527-3670				●												
Specialty Composites Delaware Industrial Park Newark, DE 19713 (302) 738-6800				●								●		●		
Stark Ceramics, Inc. P.O. Box 8880 Canton, OH 44711 (216) 488-1211												●				

TABLE 29.1 (Continued)

MANUFACTURER	PLYWOOD	GLASS FIBER	ACOUSTICAL CEILING SYSTEMS	DAMPING	SPRAY ABSORPTION	DOORS	SILENCERS/MUFFLERS	GYPSUM BOARD/PARTICLE BOARD	PLASTICS	LEAD	FOAM	CONCRETE BLOCK/CERAMICS	LEAD-LOADED VINYL/LOADED VINYL	METALS	ACOUSTICAL WALL SYSTEMS	ACOUSTICAL PANELS
Starco 1515 Fairview Avenue St. Louis MO 63132 (314) 429-5650		•						•	•							
U.S. Plywood Div. of Champion International 777 Third Avenue, New York, NY 10017 (212) 895-8000	•					•										
Veneered Metals, Inc. P.O. Box 327 Edison, NY 08817 (201) 549-3800				•											•	

30 - LONG TERM NOISE ABATEMENT

In the specification and purchase of all new items of equipment, requirements for reduced product noise should be included as part of the specification. One model noise control specification is as follows:

General

(1.1) The manufacturer shall submit sound measurements for supplied
equipment in accordance with this specification. In addition,
where the sound measurements exceed the values stated below,
the manufacturer shall advise on silencing provisions and
additional costs to meet this standard.

Measurement

(2.1) The manufacturer shall be responsible for the supplied equipment,
including any subassemblies.

(2.2) The manufacturer shall test the equipment as follows:

Item	Applicable Standard
Duct & Fan Systems	AMCA 300-67 Rest Code
Rotating Electric Machinery	IEEE 85
Pneumatic Equipment	ANSI S5.1
Machine Tools	NMTA Noise Measurement Techniques

Where no acceptable sound test exists or where manufacturer's
standard is different from the above, the manufacturer shall
submit the method of testing.

Submitting Data

(3.1) The manufacturer shall submit octave sound power data for equipment. In addition, the following information is required:

Item	Information Required
Fans	a) Total sound power level at fan casing
	b) Inlet sound power level
	c) Outlet sound power level

(Item)	(Information Required)
	d) Locations where sound measurements were taken
Rotating Machinery and Miscellaneous Devices	a) Total sound power level at the equipment casing
	b) Locations where sound measurements were taken

(3.2) If the source is highly directional, such as a cooling tower, measurements shall be submitted at several locations.

Specified Levels

(4.1) If manufacturers's data submitted in Items 3.1 and 3.2 exceeds 85 dBA, the quotation shall include the additional cost and silencing provisions necessary to meet this value.

Manufacturer's Responsibility

(5.1) It is the manufacturer's responsibility to engage an independent consultant, as required, in order to meet the noise requirements stated in this specification.

(5.2) The manufacturer shall not ship any equipment which exceeds the manufacturer's values promulgated in this specification without the purchaser's written authorization.

(5.3) If the manufacturer must perform tests at the purchaser's facility, it shall be stated in the quotation.